Environmental Mathematics
in the Classroom

This project was supported, in part, by the National Science Foundation. Opinions expressed are those of the authors and not necessarily those of the Foundation.

© 2003 by

The Mathematical Association of America (Incorporated)

Library of Congress Catalog Control Number 2002114175

ISBN 0-88385-714-6

Printed in the United States of America

Current Printing (last digit):

10 9 8 7 6 5 4 3 2 1

Environmental Mathematics
in the Classroom

Edited by

B. A. Fusaro
Florida State University

and

P. C. Kenschaft
Montclair State University

Published and Distributed by
THE MATHEMATICAL ASSOCIATION OF AMERICA

CLASSROOM RESOURCE MATERIALS

Classroom Resource Materials is intended to provide supplementary classroom material for students—laboratory exercises, projects, historical information, textbooks with unusual approaches for presenting mathematical ideas, career information, etc.

MAA Service Center
P. O. Box 91112
Washington, DC 20090-1112
1-800-331-1MAA   fax: 1-301-206-9789

Contents

Introduction

Environmental mathematics seeks to marry the most pressing challenge of our time with the most powerful technology of our time—mathematics. Environmental mathematics in the classroom does this at an elementary level. This book demonstrates a wide variety of significant environmental applications that can be explored without resorting to the calculus.

Environmental Mathematics in the Classroom includes several chapters accessible enough to be a text in a general education course, or to enrich an elementary algebra course. Ground level ozone, pollution and water use, mathematical economics, the movement of clouds over a mountain range, at least one population model, and a smorgasbord of "newspaper mathematics" can be studied at this level, and would form a stimulating course. It would prepare future teachers not only to learn basic mathematics, but to understand how they can integrate it into other topics that will intrigue their students.

Other chapters provide sufficient challenge for prospective mathematics majors. Harder population models, the spread of infections, and the survival of buffalo after the nineteenth century slaughter provide substance for such students. Those ready to apply trigonometry to real-world models also could study groundwater dissipation, the cleaning of oil spills, and the filling of oil tanks.

Environmental Mathematics in the Classroom, with only the background given in its accompanying teachers' manual, can be a text for an independent mathematics course. With the expertise of another professor, it could be the basis of an interdisciplinary course relating mathematics and science. It also could provide excellent supplementary reading for eager mathematics professors and their students, either for recreation or as a basis of independent study.

Along with providing material for fascinating undergraduate courses, *Environmental Mathematics in the Classroom* addresses two other problems: the survival of many species (including our own) and the health of the mathematical disciplines. The cheetah, for example, has a noble and ancient history. It has been admired for its grace and speed. Moreover, the cheetah is harmless except to its specific prey. Yet, the future of this superb beast is precarious because of the human degradation of its natural environment.

Mathematics too has a respected and ancient (if not as ancient!) heritage. It too has been admired for its beauty, form, and power. But this extraordinary discipline is also in a precarious state, although, unlike the cheetah, its decline cannot be entirely blamed on external circumstances.

After some decades of introspection, reveling in its intrinsic beauty and power, mathematics needs to look outward, demonstrating more publicly its usefulness. It is indeed the Queen of the Sciences but it also is the handmaiden of the sciences and the arts. Those who love math for its own sake and believe in its centrality for human survival must support those using it to clarify and solve the problems rated by young and old as most worthy of public support—environmental challenges.

This volume provides mathematicians a vehicle for addressing these challenges in their own classrooms and for helping their discipline be a worthy companion to others. It includes the terminology and concepts needed to apply mathematics to a variety of environmental fields. The instructor's manual provides supplementary background for each chapter as well as hints and answers for the exercises. Thus, learning effective applications need not be too burdensome for those whose first love is mathematics.

We have enjoyed working with the remarkable authors who have researched, developed, and written the chapters of this book. We hope you enjoy getting to know their ideas as much as we have, and we look forward to hearing your comments.

Warren Page was MAA Senior Acquisitions Editor when one of us (BAF) presented him with a list of tentative authors and titles for a book in environmental mathematics. The subject was barely in the mathematical lexicon but Warren, after a careful reading of the list, gave the project his enthusiastic approval. Andy Sterrett, editor for the Classroom Resource Materials series, took charge of the project. Andy worked with us patiently and diligently for several years, making many valuable suggestions. Elaine Pedreira and Beverly Ruedi somehow turned the authors' widely differing graphics and text formats into a coherent whole. We thank everyone involved for the necessary and valuable contributions to bringing *Environmental Mathematics in the Classroom* to the mathematical community.

B.A. Fusaro, Editor
Florida State University
Tallahassee, FL 32306-4510
fusaro@math.fsu.edu

P.C. Kenschaft, Associate Editor
Montclair State University
Upper Montclair, NJ 07043
kenschaft@math.montclair.edu

Environmental News Teaches Mathematics

Barry Schiller
Rhode Island College

Do you teach at a school where there is an opportunity to develop new courses in environmental mathematics? Is there time available in your courses to develop major projects involving building models of environmental problems? Those who can answer "yes" to such questions are fortunate indeed and this book has some great ideas for such projects and courses. Even if your answers are "no," it is possible for those of us with an interest in the environment to use this topic in a wide variety of lower division courses in such a way to teach math, develop quantitative literacy, and legitimately put environmental considerations into the curriculum.

One approach is to keep a file of clippings of items from newspapers, magazines, journals, and other information sources that relate some aspect of the environment to some aspect of mathematics. There are a huge number of possibilities, as will be illustrated. Gradually get used to adapting such items for classroom use; eventually you will make frequent use of such items. They might lead to a simple homework problem or exam question, or illustrate some point in a lecture, or help generate classroom discussions. You do not have to wait until you have the time for a big project or a new course.

I believe you cannot depend on textbooks for this. Though it is the fashion nowadays for textbooks to use real data and emphasize applications in many lower-division college math courses (and rightfully so), texts doing this may cover a wide variety of applications, so cannot often include the environment. Even when they do, it may be that the examples may not do much to encourage interest in either the environmental topic, or (for that matter) the mathematical topic. For example, if you look in the "index of applications" in one mainstream text, you will find one reference to "Recycling" that turns out to be this problem:

Let

U = set of all participants in a consumer behavior survey conducted by a national polling group.

A = set of consumers who avoid buying a product because it is not recyclable

B = consumers who use cloth rather than disposable diapers

C = consumers who boycotted a company's products because of their record on the environment

D = consumers who voluntarily recycled their garbage

The problem requests a verbal description of $A \cap C$, $A \cup D$, B complement intersection D, etc. The author could at least have used set C for the Cloth users and B for the Boycotters, to help keep the categories straight! While it is nice to raise these recycling issues, who really cares about these unions and intersections? Though students do need drill, I do not see how this problem will generate any interest in either set theory or recycling.

Another example from a competing text has one reference to "smog control" in its "index of applications":

"A new smog control device will reduce the output of sulfur oxides from automobile exhaust. It is estimated that the rate of savings to the community from the use of this device will be approximated by $S(x) = -x^2 + 4x + 8$ where $S(x)$ is the savings (in millions of dollars) after x years of use of the device. The new device cuts down on the production of sulfur oxides but it causes an increase in the production of nitrous oxides. The rate of additional cost (in millions) to the community after x years is approximated by $C(x) = 3x^2/25$."

The question goes on to ask how many years one should use this device and how much can be saved. It is good to give students practice with such problems, and help make them aware of costs, benefits (especially that there might be dollar benefits from pollution removal!) and tradeoffs, but it is evidently a pretty artificial problem. Who estimated such a formula? We know how sensitive that is to the interests of the estimators! For what interval of values of x might this be good? Not stated. Those who follow pollution issues know that cars are not a major source of sulfur dioxide. The prediction of $12 million in benefits after 2 years but a loss of $52 million after 10 years (within the lifetime of an automobile) sounds implausible.

Another problem from the second book, indexed under "pollution":

"Pollution from a factory is entering a lake. The rate of concentration of the pollutant at time t is given by $P(t) = 140t(\frac{5}{2})$ where t is the number of years since the factory started introducing pollutants into the lake. Ecologists estimate that the lake can accept a total level of pollution of 4850 units before all the fish life in the lake ends. Can the factory operate for 4 yr without killing all the fish in the lake?"

Does this problem suggest it is OK to kill 90% of the fish in the lake? The pollutant is unnamed, and the author does not even bother identifying the units of the pollutant. I doubt this can develop much interest in either the environment or in mathematics.

Contrast that example with this one from the Fall 1993 newsletter *Science for Democratic Action*, which has a regular section called "Arithmetic for Activists."

"You live one mile downwind of a uranium mill. Your trusty air monitoring equipment measured the amount of radioactivity in the air. You read 0.00037 becquerels per liter of air. Remember that 1 curie = 37 billion becquerels and that the prefix 'pico' means one trillionth. Laboratory analysis indicates this is all due to insoluble radium-228—are you above the standard?"

It is noted that the existing standard of insoluble radium-228 is .001 picocuries per liter.

This time there is no escaping the need to pay attention to units. (Though we do not mine for uranium in New England, I do like this problem. Our region is relatively dependent on nuclear energy but almost nobody pays attention to the details.) This problem is in my files, so it can be used in a "technical" math or quantitative literacy class.

I think if we are to get good real examples, it is clear that we cannot depend on textbooks, but should expect to gather them ourselves from a variety of sources.

Though even beginning liberal arts statistics courses try to cover as much statistical inference as possible, we are more likely to be training citizens in these courses than researchers. Therefore on the first day of the course I mention the most basic interpretation of the word "statistics" as a set of meaningful facts and figures. To illustrate this, I show a slide of a *Harper's Index*, a monthly compilation of interesting statistics in that sense. The students see that numbers can indeed be interesting, not only in the judgement of statisticians, but in the judgement of magazine editors who are in the business of selling magazines. Indeed the *Harper's Index* is successful enough to be a registered trademark. There is even a *Harper's Index Book* paperback! Examples from the February 1997 *Harper's Index* include:

Amount that "side agreements" in NAFTA require that the U.S. spend on environmental cleanup—$1,500,000,000

Amount the U.S. had spent by the end of 1996: 0

Older indices include:

The number of Exxon Valdez spills it would take to equal the amount of oil spilled into the Mediterranean each year: 17

Estimated percentage of the $6.7 billion spent on Superfund cleanups since 1980 that has gone to lawyers: 85

Square yards of park per inhabitant: Paris 6, New York 18

You get the idea—some amusement, but also some seriousness. If you flash a slide of an Index on a screen and ask students what is interesting about the items, they may well single out those with an environmental theme. The Indices also include numerous percentages, averages, probabilities, and ranks, but the idea is not so much to teach mathematical terminology as to encourage students to develop a lifelong belief that quantitative information is interesting and worth paying attention to.

One actually needs to clip the *Harper's Index* pages to have it available when needed. One of my colleagues who knows I use this sort of thing actually gives me a copy of the *Index* each month, and I occasionally get similar materials from other colleagues and students who know I like to collect such information. Many others use the idea of the *Harper's Index*. For example, I have a "Vital Statistics" page from the National Wildlife Federation which includes these items:

Estimated global pesticide sales in 1975: $5 billion, in 1990: $50 billion

Parts per million of DDT in human adipose tissue in the U.S. in 1970: 8, in 1983: 2

See? Do not always assume bad news!

Harper's Index has numerous references to very large numbers. This is one of the first topics developed by John Allen Paulos in his book *Innumeracy* (indeed it is referenced in the second line of the book!) because of the difficulty even educated people have in dealing with numbers in the billions, trillions etc. Think of the problem mentioned before about the picocuries. Do you have colleagues who tend to refer vaguely to zillions? It seems even newspaper headline writers do not pay adequate attention. For example, I've clipped a headline from the May 16, 1984 *Providence Journal* that says "Waste Cleanup Cost: up to $26 million." We wish it were $26 million! More than that was spent on just one superfund site in Rhode Island, the Piccillo Pig Farm. The article clearly says the cost was up to $26 billion but apparently millions and billions were all the same to the headline writer.

What we can we do to humanize the $26 billion figure? As it was supposed to be spent over a twenty-year period to clean up the sites, one could ask what it would cost on average per person, per year. One thing I like about that question is that you have to divide twice. Textbook problems illustrating the mean never seem to have such questions, even though there are numerous real situations where it applies and some students are puzzled about what to do. Another thing I like about the question is that the answer comes out so small, only about $5 per person per year. Indeed one student told me he thought it must be wrong, it was so cheap. Perhaps that is how we should argue before Congress when debating spending money on hazardous waste cleanups.

Humanizing large numbers by reducing them to a per person or per household basis as done above can be applied to a wide variety of situations. The process can also be reversed to see the cumulative impact of what sounds like insignificantly small numbers. For example, because of my interest in the impact of transportation on the environment I get a lot of information on that topic from a variety of sources. *Parking cashout* is a strategy to reduce vehicle miles by having employers who offer free parking also offer the cash value of the parking as an alternative for employees who don't use it. One parking cashout leaflet from the Conservation Law Foundation (based on a study at the UCLA School of Public Policy and Research) suggested that parking cashout can reduce auto commuting by about 625 vehicle miles per month per employee. Assuming each mile of auto use produces about .4 ounces of carbon monoxide and .038 ounces of nitrogen oxide emissions, which sound insignificant, one can ask for a reasonable estimate of the effect of how a national parking cashout program might influence the total weight of the output of these pollutants. One would need an estimate of the total labor force, and what percentage gets free parking. A ballpark estimate for the total might be of order of magnitude of a billion pounds of carbon monoxide and about 100 million pounds of the nitrogen oxides, which does sound significant.

Another example of going from a small human scale number to a large number relates to solid waste. This topic has received much attention here due to problems at the state's central landfill (literally one of the high points of Rhode Island; look for it if you ever fly into Providence) and several attempts to build solid waste incinerators which considerably alarmed people living near the proposed sites. Relate this to a problem in one of our current textbooks (in our math for elementary teachers course) asking for the surface area of a cereal box with dimensions 11 in by 2.5 in by 8 in. A check of my favorite cereal box shows that these dimensions are realistic. But instead of stopping at the answer to the textbook question (271 sq in,) why not go on to consider the surface area of a cube that would enclose the same volume? It may surprise some students that the same volume can be enclosed by only about 219 sq in of box, a savings of 52 sq in or about 19.2%. Ask for a reasonable estimate of the number of cereal boxes sold in a year (100 million households times 50 boxes a year per household was suggested) to get an estimate of the total reduction in packaging possible from redesigning cereal boxes. Ask the class why isn't it sold in the shape of a cube? Would they buy a box of cereal in the shape of a cube? What happens if the dimensions are modified to make it only somewhat more cubelike?

There must be marketing considerations here. The cereal companies want to have a large face area on which to show their brand name and logo. I would note that the cereal producers do not have to worry much about the cost of disposing of the empty boxes. The EPA indicates paper and paperboard constitute 40% of our waste stream, a considerable percentage. A newspaper report says that northeastern Governors have asked industry for voluntary cooperation in reducing packaging, though none of my students think that industry would actually pay attention to such a request.

The cereal box example makes a good multistep problem, starting just with the dimensions of the cereal box. Every math teacher knows students who can do one-step problems—they learn the procedure—but do badly on multistep problems, even if they know how to do each step. There just isn't enough practice with such problems. We math teachers have to be on the lookout for them ourselves.

That example also opens up the possibility of a percentage of percentage question. What would be the impact of a 19% reduction in 50% of the 40% of our solid waste stream? In the ideal world all students would be able to solve percentage problems, but in the world I live in, that topic is apparently not well-reinforced in the usual high school math curriculum of Algebra I, II, and Geometry. So many students need help, and practice, on percentage problems. Indeed the September 1995 *Harper's Index* notes the chances that an American 17-year-old can express 9/100 as a percentage is only 1 in 2.

Solid waste issues can lead to many percentage problems. An example I've used relates to the fact that Rhode Island, unlike our neighboring states, does not have a "bottle bill"—that is, mandated deposits on beverage containers. When that was being debated here, a flyer used by the Bottle Bill Coalition included this item that I used for this percentage problem:

"According to *Beverage Industry Magazine*, the U.S. soft drink companies spend $4.5 billion annually for packaging their product and about $800 million for ingredients. What percentage of the total of these costs is for packaging?"

Sadly, many students didn't answer this correctly, having trouble with the percentage or with the large numbers involved. Also sadly, the item didn't help pass the bottle bill, which our neighboring states have found reduces litter and increases the recycling rate.

I also have a full-page ad from the American Plastics Council entitled "Plastics. An Important Part of your Healthy Diet." I haven't decided how to actually use it, if at all. What is significant is that it has no quantitative information whatsoever.

Another example of relating human scale activities to environmental issues involves Amtrak, our intercity rail passenger service. It is facing severe financial constraints and has asked that one half cent of the gas tax be set aside for capital improvements for Amtrak. A letter to the editor from a motorist objects to their hard-earned dollars going to a service they do not use. I suggest asking for a quick guess of how much this will actually cost a typical motorist. Then ask for a procedure to come up with a more reliable estimate. For example, if you drive 15,000 miles per year in a vehicle that gets 20 mpg, the total cost of the half-cent tax would be about $3.75/year—which one student said was "nothing." Indeed it does seem like quite a bargain to keep a rail passenger system alive that might someday provide an alternative for even the most train-phobic motorists, or at least help reduce traffic on the roads for those who must drive. The point for students is, do a calculation to see what it means to you.

I've found that even environmentalists are unaware that Amtrak survival is an environmental issue. I have an article indicating the pollution in grams of hydrocarbon emission per passenger mile is about .1 for the rail mode, .2 for buses, and 2.1 for a single occupancy auto. There are similar figures for carbon monoxide. There is also fuel efficiency and all the impacts that energy extraction and transport have on the environment. Amtrak says it can carry 550 passengers one mile on only 2 gallons of fuel, while 110 five-passenger cars would take 8 gallons to do this. This could turn into various questions for a class—what does this assume about the automobile milage? If true, what is the macroscale impact?

Another source I use is the American Automobile Association, that annually publishes a summary of the costs for operating an automobile. One of these says the cost of owning and operating a new car (for 15,000 miles) is "now averaging 38.7 cents per mile.... Average per-mile cost is determined by combining operating and fixed costs... motorists nationally paid an average of 9.16 cents per mile in operating cost (gasoline, oil, maintenance, tires). Fixed costs, which include insurance, depreciation, registration, taxes and financing, average $12.14/day."

I use this in our first "quantitative business methods" course, asking students to read the article, develop the formula, and do some calculations. There is some need to be careful about units—sometimes students do not distinguish correctly between dollars and cents! A larger environmental point can be made by looking at the percentage of the total cost of driving that is the marginal or operating cost, which is only about 17%. I believe the relatively high fixed cost but low marginal cost is related to the difficulty of reducing vehicle trips, that is, it doesn't cost very much more to do a little more driving.

An example I use to illustrate functions, and marginal costs, is taken from information obtained from the New England Power Company, which generates electricity for our local utility. It has a graph of the cost (to the power company) of removing pollution from the stacks as a function of the percentage of pollution removed. Of course, it rises quite steeply after a while, especially after about 78% removal. Some students may be disappointed about what that implies for simply requiring 100% removal. Perhaps it is just as well that Congress in the 1970s did not see such graphs when they were writing the clean water laws intending to eliminate all discharges!

Energy issues are a good source of mathematical ideas. First, numerous data lists have been published, with meganumbers sometimes making interesting points. An example from our local paper is that the cost of the nuclear plant in Seabrook, NH, is enough to pay for all municipal services in Fall River for 51 years!

New York's Con Edison utility had an energy quiz in the *New York Times*. A sample question follows. (Do you know the answer?)

Burning the oil required to light one ordinary 75-watt bulb for a year releases how many pounds of gases that might contribute to environmental problems?

(a) 275 pounds

(b) 5 pounds

(c) 12 pounds

(d) 1 pound.

Con Ed said the answer is (a). A less benign example from a nearby gas utility is its rate structure. A handout given to customers describes this monthly commercial rate structure:

"Customer Charge is $8.15. First 4000 ccf 63.57 cents per ccf. Over 4000 ccf: 55.76 cents per ccf."

This gives math teachers the opportunity to discuss not only a real piecewise defined function (writing it algebraically poses more of a challenge for students than I had anticipated) but also marginal costs and the effect of a declining rate structure on the conservation ethic.

Another energy item from the *New York Times* relates to power line electromagnetic fields and cancer. The headline reads "Federal Panel Says Electric Fields Pose No Known Hazards" but the article itself says that

> "the statistically weak link between leukemia and proximity of large power lines may be due to unknown factors with no connection to electromagnetic fields. Possible outside factors which need to be looked into more closely include the age of homes and their construction features, pollution, local air quality and heavy traffic near power lines . . . "

This can be a good springboard for a discussion of confounding in statistics. I can report that despite the headline, the article did not encourage my students to locate near a power line.

Being on the lookout for examples of confounding in an environmental context can be rewarding. For example, the students see the point instantly in a *New York Times* "Week in Review" article headlined "Reading at 55 Miles Per Hour," which reports that accidents were 41% higher in billboard areas. It would be nice to use this to justify removing the billboards and so improving the scenery, but it can be that advertisers prefer to locate billboards where there is heavy traffic, and thus more accidents.

Population growth models are a staple of environmental mathematics and there are plenty of graphs from environmental groups showing exponential growth. A more unusual graph is from the 11/17/96 *New York Times* headed "The Population Explosion Slows Down." The "today" point on the graph is just where the rate of increase starts dropping. It came just at the right time for my calculus class studying the second derivative. When I put up a slide of this graph they immediately saw it as a point of inflection. However, anyone using this beware: the slowdown in growth is only a projection!

Do not think always being on the lookout for examples that can be used in classes will make it too hard to just relax and read the paper. I believe it is worth some effort to be able to use a wide variety of topics: Reading that Sarawak (in Malaysia) had 21000 square miles of rain forest but it was being destroyed at a rate of 1000 square miles per year became both a linear equation problem and a comment about the destruction of the tropical rain forest; You do not have to make up hypothetical probability distributions— you, like my students, may be surprised that the city of Providence reports that 23% of their households have no cars at all, a group usually forgotten about by local transportation planners with cars. The entire distribution for 0, 1, 2, 3, 4, 5 vehicles is .23, .42, .27, .06, .016, .004 respectively; The cost C per bus-mile for a bus system is given by the function $C = 0.88 + 27.04/S + 23.874/U$, where S = average operating speed, U = peak vehicle utilization. Note what happens as $S \to 0$. Gridlock!

Roundoff can be important! An environmental group commenting on a proposed ozone air quality standard of 80 parts per billion warns not to allow roundoff (to 80 ppb) from actual pollution levels up to 85 ppb to meet the standard. In other words, they want the 0 in 80 to be a significant digit so that a reading of 80.5 ppb would be a violation to be addressed. They say, "Rounding up means people have to breathe more pollution in the air."

This poll result in the *New York Times* can illustrate how newspapers may report the margin of error in surveys: 20% favor reducing spending on the environment, 74% say that is unacceptable. The *Times* explains that in 19 cases out of 20, samples of the size used would result in a margin of error of no more than 3% either way. We get an interpretation of 95% confidence intervals while indicating something about public support for the environment.

Teaching at a state college where most students are Rhode Islanders leads to looking for a local angle. Perhaps attention to the nearby is a good idea everywhere. It is easy to get data, graphs, and information on such local environmental issues as the depletion of fish off our coast, pollution of Narragansett Bay, and transport of air pollution.

What do students think of all this? I would like to live in a world where my ideas get universally favorable responses (some colleagues seem to live in that kind of world) but I think a more accurate summation is obtained by this quote from my student evaluations: "Instructor tried hard to bring interesting side issues into mathematics, sometimes successfully."

By now, my files of such articles are quite large. I would be glad to share any of them with colleagues who write to me at Rhode Island College. I believe that such files can be used to encourage students to maintain a lifelong interest in paying careful attention to quantitative information, especially with regard to the environment.

This approach suggests exercises that make both environmental and mathematical points. A sample of such exercises follows.

EXERCISES

1. What is a verbal description of set $A \cup D$ for the sets described on page 2. Do you think it is likely that $D \subseteq A$? Explain.

2. Actually answer, if possible, the question about whether or not you meet the insoluble radium-228 standard as asked on page 3.

3. *Harper's Index* recently noted this: Estimated amount of gas wasted in U.S. traffic jams each day, in gallons: 12,600,000. Assuming this, give a reasonable estimate for the cost of wasted gas to American motorists per year.

4. Examples of the need to humanize large numbers:
   (a) Is the United States (land area about 3.54 million square miles, population 272.6 million) more densely populated than the rest of the world (land area 50.6 million square miles, population 6.00 billion)? Explain carefully (1999 data).
   (b) Based on the following, over the period 1980 to 1997, which increased faster, population or the quantity of plastics in municipal solid waste? Explain clearly.

Year	total municipal solid waste	% plastics	U.S. population
1980	303 billion pounds	5.2	227.7 million
1997	434 billion pounds	9.9	268.0 million

(c) What was the change in the amount of plastic waste generated per person per day?

5. With regard to the shape of packaging, compare the surface area of packaging of a product packaged in cylinders of radius 4 in, height 4 in, with a cubical package that contains the same volume.

6. Use the AAA data given for the average cost of driving to answer the following.

   (a) How much does it cost the average motorist to drive another 100 miles?

   (b) Compare the cost per mile for the driver who does 15,000 miles in a year with the driver who does 30,000 miles.

   (c) Similarly use the data given about the commercial cost of natural gas to compare the average cost per ccf for a customer who uses 3500 ccf in a month with those who use 7000 ccf in a month.

Using Real-World Data to Understand Environmental Challenges

Iris B. Fetta

Clemson University, SC

Real-world information collected in numerical form is called *data*, and many topics relevant to our lives can be investigated by using data. One of the most important of these is the environment. In the examples and activities that follow, you will be using real-world data to study mathematical concepts while increasing your environmental awareness.

Data usually represent only a partial view of a real-world situation. For example, the individual data points do not indicate what happened between two particular data values or what might happen just outside the limits of the data. Such predictions are often possible and meaningful in the context of a real-world situation when we use functions to describe the pattern exhibited by the data. Also, when functions are used, we can apply mathematical concepts and techniques while giving meaning to the variables that describe the relationships shown by the data. Realize that the process of fitting curves to data is complex. It is not our purpose to attempt to teach that process here. Instead, we use familiarity with basic function shapes as a method of motivating interest in mathematical applications for environmental topics.

Communication of your ideas and results is necessary if your work is to be meaningful to anyone else. Whenever using functions in an applied setting, you should give a clear description of what your variables represent, how they are measured, and their relationship to each other. In order to be useful to you or anyone else, this description should contain symbols for the function *input* (the independent variable), the function *output* (the dependent variable), the equation relating the input and output, units of measure whenever they are applicable in the problem situation, and descriptions of both the input and output. We choose to call such a description a *model*.

POLYNOMIAL FUNCTIONS

The Environmental Protection Agency (EPA) is the principal Federal agency responsible for pollution abatement and control activities. Based on reports[1] from nearly 23,000 manufacturing facilities that meet established thresholds for manufacturing and processing using the EPA's list of more than 300 chemicals covered, the data in Table 1 give information on the release (primarily into air and water) of toxic chemicals between 1989 and 1993.

TABLE 1   Release of Toxic Chemicals by Manufacturing Facilities

	Toxic chemicals released (millions of pounds)				
Year	1989	1990	1991	1992	1993
Primary metal industries	522.9	476.7	424.7	348.6	328.6
Chemical and allied products industries	2093.3	1629.5	1546.9	1543.0	1308.4

A graph of data, called a *scatter plot*, provides much information about the shape of the graph of the function we use to describe the pattern of the data. Whenever the input for the function (in this case, the year in which the data were collected or reported) involves large numbers, some or all of the coefficients of any function fit to the data will probably also be large. Thus, to make it easier to deal with such data, we shift the points to the left. This can be accomplished in several ways, but we choose to shift the data so that 1989 corresponds to 0. That is, instead of the input variable representing the year in which the data were collected, we choose to let it represent the number of years since 1989. The shifted data appear in Table 2.

TABLE 2   Toxic Chemicals Data with Shifted Input

	Toxic chemicals released (millions of pounds)				
Years since 1989	0	1	2	3	4
Primary metal industries	522.9	476.7	424.7	348.6	328.6
Chemical and allied products industries	2093.3	1629.5	1546.9	1543.0	1308.4

We first consider the toxic chemicals released from primary metal industries. A scatter plot of the shifted data appears in Figure 1.

---

[1]U.S. Bureau of the Census, *Statistical Abstract of the United States: 1995* (115th edition), Washington, DC, 1995.

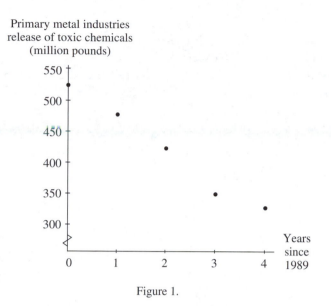

Figure 1.

What pattern do these data show? Other than the 1992 data point, the shape is basically a straight line. In fact, if you fit both linear and quadratic functions to these points, there is not much difference in how their graphs fit the data between the years 1989 and 1993. Desiring to keep the function as simple as possible, we use the linear model

$$m(x) = -51.67x + 523.64 \text{ million pounds}$$

of primary metal industry toxic chemicals released, where x is the number of years after 1989. A graph of this function on the scatter plot appears in Figure 2.

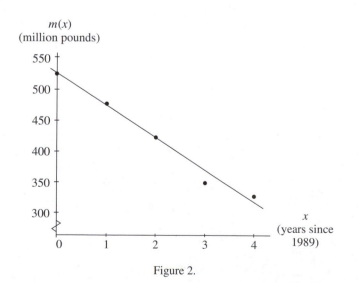

Figure 2.

Notice that if the data were shifted other than 1989 years to the left of its original position, you would have a different, but equivalent, model. It is therefore very important to describe what your input variable represents.

We next consider the largest amount of pollutants from industries, the toxic chemicals released from chemical and allied products facilities (see Figure 3).

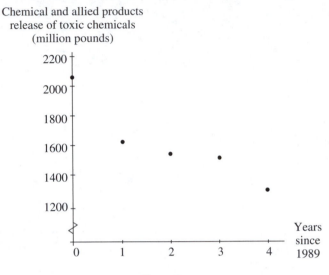

Figure 3.

What pattern does this scatter plot of the shifted data show? Note the different way the data "bend" before and after 1991. This shape should be familiar because it is that of a

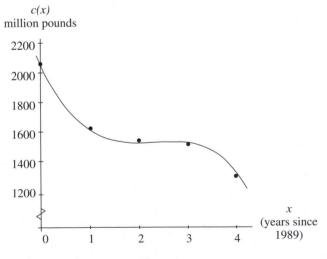

Figure 4.

cubic function. We find the cubic model to be

$$c(x) = -50.9917x^3 + 344.3143x^2 - 757.6155x + 2093.3986 \text{ million pounds}$$

of toxic releases from chemical and allied products where x is the number of years after 1989. A graph of the equation on the scatter plot appears in Figure 4.

Once you have found a model for the data, you can use your technology to obtain reasonable estimates of answers to questions such as, "When between 1991 and 1993 were toxic releases from chemical and allied products industries the most? the least?" and "In what year were 1,600,000,000 pounds of toxic chemicals released?"

General guidelines for fitting basic polynomial functions to data points are summarized by the following:

Function	Equation	Shape of Data
linear	$y = ax + b$	Straight line pattern, no or very little bending (curvature) shown by data
quadratic	$y = ax^2 + bx + c$	Single curvature, possible high or low point, but not both, shown by data
cubic	$y = ax^3 + bx^2 + cx + d$	Data bends or curves in two different ways and has 0 or 2 turning points

When observing the shape of data points, you should always construct the scatter plot so that the data points "fill" the view used for the graph. For instance, when scatter plots are drawn with technology, be sure to use an autoscaling feature that is designed to give the best view of the data. Data points that are observed using inappropriate horizontal and vertical views often can have misleading shapes. This could result in the choice of an inappropriate curve to fit the data. For instance, if the scatter plot shown in Figure 4 were drawn using much wider horizontal and vertical views than those indicted by the data, a plot similar to the one in Figure 5 could result.

The data points shown in Figure 5 are not spread out over the entire view used for the graph. Thus, this scatter plot does not exhibit the marked change in curvature leading to the choice of a cubic function.

AVERAGE RATES OF CHANGE

Consider how the outputs of the function shown in Figure 4 change between successive years. For instance, between 1990 and 1991, the amount of toxic chemicals released from chemical and allied products industries decreased by about 81.6 million pounds because $c(2) - c(1) \approx -81.6$. Figure 6 shows a line passing through the points $(1, c(1))$ and $(2, c(2))$. Such a line is called a *secant line*. The secant line shown in Figure 6 has a slope of approximately -81.6.

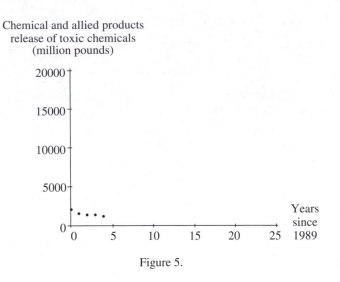

Figure 5.

The slope of a secant line is also called the *average rate of change* of the quantity that represents the output. The units on the average rate of change are

$$\frac{\text{(output units)}}{\text{(input units)}}$$

because slope of a line is calculated as rise/run. We can interpret the average rate of change shown graphically in Figure 6 by saying, "Toxic releases from chemical and allied products industries were decreasing, on average, by 81.6 million pounds per year between 1990 and 1991."

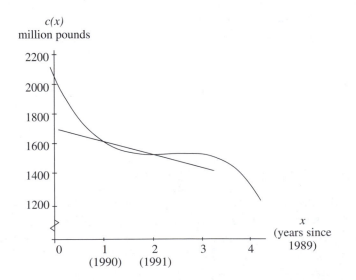

Figure 6.

On average, between which two successive years were the toxic releases from chemical and allied products industries decreasing most rapidly? To answer this question, consider Table 3, which shows the slopes of secant lines joining outputs calculated from the cubic function by using successive year inputs.

TABLE 3   Slopes of Secant Lines Calculated Using the Cubic Function for Toxic Releases from Chemical and Allied Products Industries

Years	Secant Line Slope	Average Rate of Change (millions of pounds per year)
1989–1990	$\dfrac{c(1) - c(0)}{1 - 0} \approx -464.3$	−464.3
1990–1991	$\dfrac{c(2) - c(1)}{2 - 1} \approx -81.6$	−81.6
1991–1992	$\dfrac{c(3) - c(2)}{3 - 2} \approx -4.9$	−4.9
1992–1993	$\dfrac{c(4) - c(3)}{4 - 3} \approx -234.1$	−234.1

We see that the toxic releases from chemical and allied products industries were decreasing the fastest between 1989 and 1990. On average, toxic releases were decreasing by about 464.3 million pounds per year between 1989 and 1990.

Average rates of change can also be calculated using data points without fitting a curve to the data. The output values calculated from the function and those from the data should be very close if the curve fits the data well. When you need an average rate of change between two values and you do not have one or more corresponding data points, it is then necessary to fit a curve to the data and use the function outputs for your calculations.

EXPONENTIAL MODELS

Since 1850, the National Park Service has administered our National Parks, National Monuments, National Seashores, National Historical and Military areas, and National Parkways. Data[2] on visits to these areas collected by the National Park Service are given in Table 4.

TABLE 4   Visits to National Parks, Monuments, and Allied Areas

Year	1914	1924	1934	1944	1954	1964	1974
Total visits (millions)	0.24	1.67	6.34	8.34	54.21	111.39	217.40

[2]George T. Kurian, Ed., *Datapedia of the United States, 1790–2000*, Bernan Press, 1994.

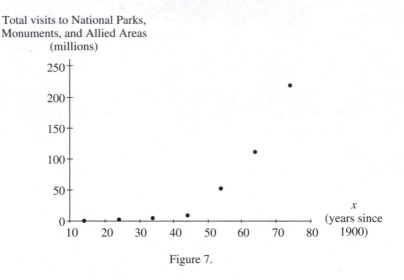

Figure 7.

The scatter plot of these data, shown in Figure 7, indicates a curvature but no obvious change in the way the data curves or bends.

You might think that a quadratic function would be appropriate. However, notice that the pattern shown by the data indicates that the number of visits started off fairly slow and then increased very quickly. This type of change is what happens with *exponential growth*. Figure 8 shows both quadratic and exponential functions fitted to the total visits data. These models are

quadratic model: $T(x) = 0.1040x^2 - 5.8663x + 72.3720$ million visits x years
                      since 1900

exponential model: $V(x) = 0.0891(1.1170^x)$ million visits x years since 1900

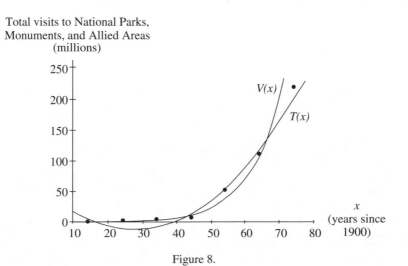

Figure 8.

If you prefer, by using a little algebra $y = V(x)$ can be written in the form $y = a(e^{kx})$:

$$V(x) = 0.0891(1.1170^x)$$
$$= 0.0891[e^{\ln(1.1170)}]^x$$
$$= 0.0891[e^{0.1106}]^x$$

Thus, an equivalent function is $V(x) = 0.0891(e^{0.1106x})$ million visits x years since 1900.

Which function should be fit to the total visits data? If you are concerned with the fit for all years between 1914 and 1974, the exponential model is more appropriate because the number of total visits cannot be a negative number. (It also fits the data very well for 1914–1944.) However, the quadratic model seems to be a better fit for the years 1944 through 1974. Predicting the number of total visits beyond 1974 with either function should be viewed with caution. Obtaining more data and observing the resulting pattern should be your next step if current-day observations about the number of total visits is desired.

General guidelines for fitting exponential functions to data points are summarized as follows:

Function	Equation	Shape of Data
exponential	$y = a(b^x)$ $y = a(e^{kx})$	Data show no change in curvature, no turning points, and outputs increase slowly and then rapidly rise; as the input gets smaller and smaller, the output gets closer and closer to a constant value
		or outputs start off high and decrease rapidly; as the input gets larger and larger, the output becomes closer and closer to a constant value

Caution: Shifting the data so that smaller input values result is often optional when finding polynomial models, but it is essential when using most technologies to find exponential models. If you do not use small input values when finding an exponential model, a meaningless model or an error message will likely be the result.

PIECEWISE MODELS

From personal need to industrial and farming uses, water is an essential element of our daily existence. As the U.S. population has grown, so has the amount of water used by its inhabitants. Listed in Table 5 is the estimated U.S. daily average water use,[3] in billions of gallons.

The scatter plot of these data, shown in Figure 9 with the input shifted to be the number of years after 1900, indicates two distinct patterns. Between 1900 and 1960, you

---

[3] *Ibid.*

TABLE 5   *Estimated Daily Water Use in the United States, 1900–1985*

Year	1900	1910	1920	1930	1940	1950
Daily water use (billion gallons)	40.2	66.4	91.5	110.5	136.4	202.7
Year	1960	1965	1970	1975	1980	1985
Daily water use (billion gallons)	322.9	269.6	327.3	420.0	450.0	400.0

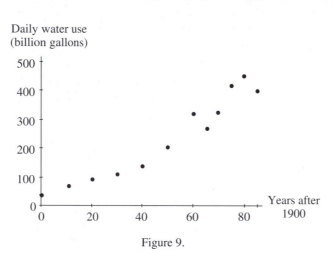

Figure 9.

might consider fitting a quadratic or an exponential curve. However, if you look closely at the scatter plot, the data appear to bend down between 1900 and 1930 and bend up between 1930 and 1960. So, a cubic function is an appropriate choice. The pattern between 1960 and 1985 is more easily identified as indicating the shape of a cubic function.

The function that best fits the 1900–1985 data is therefore composed of two distinct curves and is called a *piecewise function*. A piecewise model for the 1900-1985 daily estimated water use data is

$$w(t) = \begin{cases} 0.0028t^3 - 0.1716t^2 + 4.8746t + 37.9143 & \text{when } 0 \le t \le 60 \\ -0.0925t^3 + 10.0409t^2 - 1430.5862t + 33{,}978.6643 & \text{when } 60 < t \le 85 \end{cases}$$

billion gallons of water, where t is the number of years after 1900.

Notice that even though $t = 60$ is an input value common to both data sets, you should not include this value in both pieces of the model because the output values may be different. (Remember that a function can have no more than one output for each input.) This is, in fact, the case for the daily water use equation because the output from the left piece at $t = 60$ is about 321.5, while the output from the right piece is approximately 321.2. Because the output of the left piece is closer to the actual data value, 322.9, the input of 60 is included with the first piece. Figure 10 shows the piecewise function $w(t)$ graphed on the scatter plot of the data.

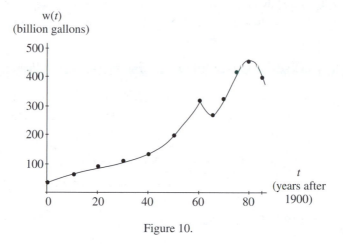

Figure 10.

Why did water use increase for so many decades and then begin to fluctuate so dramatically? Could it have had anything to do with steam-powered engines? Is it due to the decrease in farmland in the United States? Is it possibly due to changing weather patterns? Are there other reasons, such as conservation efforts? Data on related topics might help you conjecture some answers to these questions. Water is a very important natural resource. The data shown in Table 5 and in Figures 9 and 10 confirms its importance in our lives. How are pollutants affecting our water supply? It is important to find the answers to these questions and use patterns from the past to help us in the fight to conserve our natural resources.

EXERCISES

1. **(a)** Release of toxic chemicals by all manufacturing facilities[4] amounted to 2.8 billion pounds in 1993, down 25% from 1990. Use this information to find the amount of toxic chemicals released by all manufacturing facilities in 1990.

   (b) Figures 1 and 2 show that the toxic chemicals released by industries in the U.S. are decreasing. Find, in the library or one of your other textbooks, some possible reasons as to why this decrease occurred.

   (c) Consider again the model for release of toxic chemicals from primary metal industries between 1989 and 1993:

   $$m(x) = -51.67x + 523.64 \text{ million pounds}$$

   where x is the number of years after 1989.

---

[4]U.S. Bureau of the Census, *Statistical Abstract of the United States: 1995* (115th edition), Washington, DC, 1995.

(i) Do negative values of x have meaning with this model? Explain.

(ii) Find the meanings of the words *interpolation* and *extrapolation*.

(iii) Use the given function to estimate the amount of toxic chemicals released from primary metal industries in 1975, 1990, and 2000. Do you feel these results are valid? Discuss.

2. Chlorofluorocarbons (CFCs 11 and 12) are greenhouse gases that result from the use of refrigeration, air conditioning, aerosols, foams, solvents, and so forth. CFCs are expected to contribute about 20% of projected global warming, and their life span in the Earth's atmosphere is between 65 and 110 years. Table 6 shows estimated atmospheric release[5] of CFC-12 between 1958 and 1976.

TABLE 6   *Atmospheric Release of CFC-12, 1958–1974*

Year	1958	1960	1962	1964	1966	1968	1970	1972	1974
CFC-12 (millions of kilograms)	66.9	89.1	114.5	155.5	195.0	246.5	299.9	349.9	418.6

(a) What are greenhouse gases? What is the *greenhouse effect* and why should it concern you?

(b) Observe a scatter plot of the data, and fit a function to the data points.

(c) Use both the data and your function to find the average rate of change of the atmospheric release of CFC-12 between 1958 and 1974. Do either of your results indicate the changing nature of the data? Explain.

(d) Use your function to estimate the atmospheric release of CFC-12 in 1976. Under what condition would this prediction be valid?

3. At least 60% of the countries submitting national reports to a preparatory United Nations meeting preceding the Earth Summit conference in Brazil in 1992 said that solid wastes were among their biggest environmental concerns. Citizens of the United States and Canada generate twice as much garbage per person as individuals in other industrialized countries.[6] Table 7 gives the per capita solid waste,[7] in pounds per day, for the United States.

TABLE 7   *How U.S. Per Capita Garbage Is Changing*

Year	1960	1970	1980	1990	1995
Total waste generated (pounds per day)	2.66	3.27	3.61	4.00	4.21

---

[5]Ronald Bailey, Ed., *The True State of the Planet*, The Free Press for the Competitive Enterprise Institute, 1995.
[6]World Resources Institute, *The 1993 Information Please Environmental Almanac*, Houghton Mifflin Company, 1993.
[7]*Ibid.*

(a) Discuss the meaning of the words *per capita*.

(b) According to the EPA, the amount of solid waste that the average U.S. resident produces is well over one-half ton a year. Based on the 1990 data, how accurate is this statement?

(c) Draw a scatter plot and discuss the behavior of the data.

(d) Find a model to fit the data.

(e) Use your function to predict the per capita waste generated in the year 2000. The *1993 Information Please Environmental Almanac* projects that the per capita garbage in 2000 will be 4.41 pounds per day. How close to this value is your prediction?

(f) Give some reasons as to why you should be concerned about the data presented in Table 7.

4. Even though U.S. per capita total waste generated is increasing, food waste is decreasing. The U.S. per capita food waste between 1960 and 1995 can be modeled[8] by

$$w(t) = -0.0026t + 0.3687 \text{ pounds per day}$$

where t is the number of years since 1960.

(a) Interpret the meaning of $w(35) \approx 0.28$.

(b) Draw a graph of w. Write a sentence interpreting the w-intercept in the context of this application.

(c) Find the t-intercept. Do you feel the information presented by this value is meaningful? Explain.

(d) What information about per capita food wastes is given by the slope of the graph of w?

5. Nitrogen oxide emissions from automobiles and trucks contribute to smog and acid rain. The most important constituent of smog is ozone, a gas that is toxic to most living organisms and often causes eye irritation and respiratory problems. Nitrogen oxides in the atmosphere reduce potential crop yields and contribute to acid rain and global warming. Table 8 shows the total yearly emissions,[9] in millions of metric tons, of nitrogen oxides in the United States between 1940 and 1990.

TABLE 8 NO_x Emissions in the United States, 1940–1990

Year	1940	1950	1960	1970	1980	1990
NO_x emissions (millions of metric tons)	6.9	9.4	13.0	18.5	20.9	19.6

(a) Discuss the behavior of the data. Based on your observations, fit a curve to the data.

---

[8] *Ibid.*

[9] U.S. Bureau of the Census, *Statistical Abstract of the United States: 1992* (112th edition), Washington, DC, 1992.

(b) Use your function to predict the level of emissions in the current year. Do you feel your prediction is accurate? (Find information on the Clean Air Act Amendments and emission control standards on automobiles before answering.)

(c) Use your function to estimate the year (between 1940 and 1990) when nitrogen oxide emissions were the greatest. How does the additional information you found in part (b) relate to this result? Estimate the maximum emissions.

(d) Use your function to estimate the year (between 1940 and 1990) when nitrogen oxide emissions were the least. Estimate the amount that year.

(e) Between which two successive ten-year periods (beginning with 1940 and ending with 1990) was the average rate of change of nitrogen oxide emissions the greatest? Find and interpret the average rate of change at that time.

(f) Would you expect other air pollutants to exhibit similar behavior over the same time period? Explain. Find data on other toxic emissions and fit curves to check your conjectures.

6. The 1980–1992 government expenditures and total expenditures from all sources[10] for pollution abatement and control, in constant 1987 dollars, are listed in Table 9.

(a) What is the meaning of the phrase "in constant 1987 dollars?" Why do you feel the data in this activity are given in constant dollars?

TABLE 9   Pollution Abatement and Control Expenditures, 1980–1992 (in constant 1987 dollars)

Year	Government Expenditure (billions of dollars)	Total Expenditure (billions of dollars)
1980	15.3	65.6
1981	13.5	63.6
1982	13.2	61.7
1983	13.0	63.8
1984	14.1	68.9
1985	15.0	72.8
1986	16.3	77.5
1987	17.8	77.6
1988	18.2	81.5
1989	18.9	81.6
1990	20.2	83.9
1991	20.6	83.3
1992	21.3	87.6

---

[10]U.S. Bureau of the Census, *Statistical Abstract of the United States: 1995* (115th edition), Washington, DC, 1995.

(b) Find a cubic model for government expenditures as a function of the years since 1980. Discuss the fit of the curve to the data.

(c) Find a cubic model for total expenditures as a function of the years since 1980. How well does the curve fit the data?

(d) Name some sources, other than government spending, for funds used for pollution and abatement control.

(e) Use your functions from parts (a) and (b) to find a model for all expenditures, other than from the government, on pollution control and abatement between 1980 and 1992. Discuss why it is necessary to have the same input for the two functions found in parts (a) and (b) in order to find a meaningful model in this part of the activity.

(f) Use the function found in part (e) to estimate funds for pollution abatement and control from sources other than government in 1985. The actual amount spent from other sources was $56,949,000,000. How close was your estimate?

7. Chlorofluorocarbons (CFCs) were banned from use in aerosols in the United States in the 1970s, and the atmospheric release of CFCs declined. Unfortunately, the decline was temporary due to increased use of CFC-emitting substances in other countries. In an international effort to reduce CFC emissions, the Montreal Protocol was ratified in 1987. Table 10 shows the atmospheric release[11] of CFC-12, one of the two most prominent CFCs.

TABLE 10   Atmospheric Release of CFC-12, 1974–1992

Year	1974	1976	1978	1980	1982	1983	1986	1988	1990	1992
CFC-12 (millions of kilograms)	418.6	390.4	341.3	332.5	337.4	359.4	376.5	392.8	310.5	255.3

(a) How evident is the information given at the beginning of this activity when you observe the behavior of the data in Table 10?

(b) Group the data in two subsets, each having one common data point. Fit a function to each set of data, and combine the two functions to form a piecewise model for the 1974–1992 data.

(c) Use the function found in part (b) to estimate when, between 1974 and 1992, atmospheric release of CFC-12 was the greatest and the least. What are the maximum and minimum amounts?

(d) Does your function predict a time when there will not be any atmospheric release of CFC-12? If so, what is that date? Is this a realistic prediction? Give reasons to support your conclusion.

---

[11] Ronald Bailey, Ed., *The True State of the Planet*, The Free Press for the Competitive Enterprise Institute, 1995.

(e) Combine your result to part (b) with the data and function from Exercise 2 to form a piecewise model for the 1958–1992 data. Sketch, on graph paper, a graph of the data and the three-part piecewise function.

8. (a) Compare your result in part (d) of Exercise 2 to the data given in Exercise 7. Discuss the hazards of predicting output values outside the range of the data used to find the function.

(b) Use the Internet to find more recent data (after 1992) on the atmospheric release of CFC-12. Add these data points to your scatter plot of the data given in Exercises 2 and 7. Find an appropriate piecewise function for all the data.

(c) Prepare a 5–10 minute oral presentation on the effects of CFCs in our environment along with U.S. and world efforts to help solve this problem. Your report should include the use of appropriate models and graphs to support your conclusions.

9. The resident population of California,[12] measured in people per square mile of land area, is given in Table 11.

TABLE 11   Density of Resident Population of California, 1850–1990

Year	Population Density (people per square mile of land area)
1850	0.6
1860	2.4
1870	3.6
1880	5.5
1890	7.8
1900	9.5
1910	15.3
1920	22.0
1930	36.2
1940	44.1
1950	67.5
1960	100.4
1970	127.6
1980	151.7
1990	190.8

(a) Why might an exponential function be appropriate to fit these data? Fit an exponential function to the data. Comment on how well the function fits the data points.

---

[12]George T. Kurian, Ed., *Datapedia of the United States, 1790–2000*, Bernan Press, 1994.

(b) Is there another function that could be fit to these data? If so, give reasons for your choice and find the model.

(c) Use your functions from parts (b) and (c) to predict the resident population density of California in the year 1912. Which prediction do you feel is better and why?

10. Table 12 gives yearly emissions of carbon monoxide,[13] in millions of metric tons, in the United States.

TABLE 12 Emissions of Carbon Monoxide in the United States, 1980–1990

Year	Carbon Monoxide Emissions (in millions of metric tons)
1980	117.0
1981	111.6
1982	105.4
1983	105.2
1984	102.5
1985	97.9
1986	95.2
1987	90.1
1988	89.9
1989	84.7
1990	83.8

(a) Name some sources of carbon monoxide emissions. Has there been any legislation during the past two decades that would cause a decline in pollution due to carbon monoxide emissions?

(b) Draw a scatter plot of the data. Fit linear, quadratic, and exponential functions to the data. Discuss the appropriateness of each function.

(c) Find the average rate of change of carbon monoxide emissions between 1980 and 1990 using the data and each of the functions found in part (b). Compare and constrast these values.

(d) Predict, using each of the functions found in part (b), carbon monoxide emissions in 1991 and 1992. The actual carbon monoxide emissions in 1991 were 82.3 million metric tons, and carbon monoxide emissions in 1992 were 79.1 million metric tons. Compare these values to your predicted values.

(e) Graph the three functions found in part (b) between the years 1990 and 2050. Discuss the future behavior of all three functions. What do you predict will happen to

---

[13]U.S. Bureau of the Census, *Statistical Abstract of the United States: 1994* (114th edition), Washington, DC, 1994.

carbon monoxide emissions in the future? Do any of these three functions support your conjecture?

REFERENCES

Efron, B., "Computer-Intensive Methods in Statistical Regression," *Siam Review*, Volume 30, Number 3, September 1988.

Fetta, I., *Graphing Calculator Instruction Guide* to accompany *Calculus Concepts: An Informal Approach to the Mathematics of Change*, Houghton Mifflin, 2002.

LaTorre, D., J. Kenelly, I. Fetta, L. Carpenter, C. Harris, *Calculus Concepts: An Informal Approach to the Mathematics of Change*, Houghton Mifflin, 2002.

Skala, H., "Will the Real Best Fit Curve Please Stand Up?," Classroom Computer Capsule, *The College Mathematics Journal*, Volume 27, Number 3, May 1996.

The Price of Power

Christopher Schaufele
Nancy Zumoff[1]
Kennesaw State University, GA

INTRODUCTION

A coal burning power plant located 3 kilometers from a small lake known as Lonesome Lake is put into operation to provide a small village with electricity. Considerable amounts of sulfur (S) are locked up in coal and when coal is burned, the sulfur is released into the air in the form of sulfur dioxide (SO_2). At this time the SO_2 acquires an additional oxygen molecule to become sulfur trioxide (SO_3). Rain or snow absorbs SO_3 and produces sulfuric acid (H_2SO_4), which is deposited along with the precipitation onto the ground and into streams. The presence of excessive amounts of sulfur in the atmosphere means more acid falls with the precipitation. When this happens, the precipitation is called *acid rain*. Acid rain not only harms vegetation, but when rivers and lakes become too acid, fish and other aquatic life die out and drinking water obtained from these sources becomes foul.

This study examines the effect of sulfur emissions on the water of Lonesome Lake. To do this, we look at the annual emission of sulfur from the power plant together with the annual precipitation. The chemical reactions of the sulfur, oxygen, and water create hydrogen ions, which determine the acidity of a liquid. The acidity is measured numerically by a pH scale; a formal definition is given in Section 2. The greater the amount of sulfur, the greater the number of hydrogen ions; this makes the liquid more acid. A low pH value indicates acid liquid, whereas a higher pH value shows alkaline, or less acid, liquid. A pH value of 7 means the water is neutral, neither acid nor alkaline. The goal of this chapter

---

[1]This work is supported by U.S. Dept. of Education, FIPSE, grant #P11A30555 and National Science Foundation grant #DUE-9354647.

is to connect the pH of the water in the lake and the effect this might have on its fish population.

1   SULFUR DIOXIDE EMISSION

Not all the sulfur emitted from the power plant ends up over the watershed of the Nizhoni River, which feeds the lake. The first step in deciding the effect of sulfur emission on Lonesome Lake is to determine the percentage of sulfur that actually does wind up over its watershed. Figure 1 shows the relative position of the power plant, lake, and watershed.

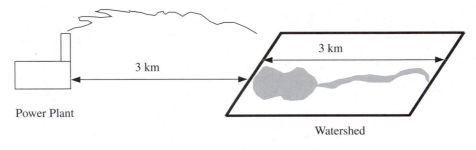

Figure 1.

Dispersion of the SO_2 is dependent upon weather conditions, emission rate, and the effective height of the smokestack. We assume a constant emission rate and a fixed weather pattern with the lake situated downwind from the power plant. The effective height of the smokestack is the distance from the ground to the horizontal centerline of the smoke plume (see Figure 2).

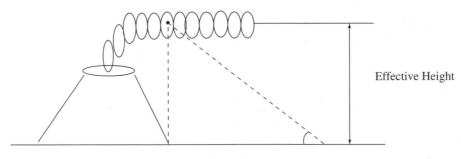

Figure 2.

The height is determined using right triangles. If we stand at a known horizontal distance from a point directly underneath the plume and measure the angle between the horizontal and a line to the centerline of the plume, this angle is called the *angle of elevation*.

In the following exercises, you will learn much more by working in groups.

Exercise 1. If the angle of elevation of the centerline of the smoke plume is 50.2° from a point 100 meters from the base of the smokestack, find the effective height of the plume (round to the nearest 10 meters).

Exercise 2. The annual emission of sulfur dioxide from the plant is $10^{7.81}$ grams. The emission rate of SO_2, measured in grams per second, is assumed to be constant. Determine the emission rate and denote it by Q. Use these conversion factors and round your answer to 4 decimal places:

1 year = 365 days;
1 day = 24 hours;
1 hour = 3600 seconds.

The ground-level concentration of the sulfur dioxide varies with the distance downwind from the stack. The steps below lead to the derivation of a function that estimates this concentration at a specified distance downwind. This function is used to approximate the percentage of sulfur emissions that lie over the watershed. As might be expected, the ground-level SO_2 concentration increases from the point of emission to its maximum, then drops off rather rapidly and almost levels out as it approaches 0 at a distance (see Figure 3.)

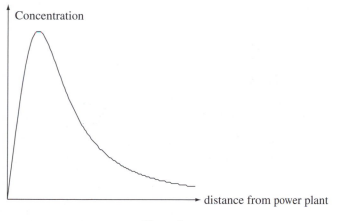

Figure 3.

Even though environmental engineers use a more sophisticated function called a *Gaussian distribution* to measure concentration, such a pattern can be approximated by a rational function with a denominator of degree larger than that of its numerator. In this case, a function of the form

$$C(x) = \frac{Q}{u} \left[\frac{Ax}{x^3 + B} \right]$$

will give a reasonable approximation. Here,

x = distance downwind from the stack,

$C(x)$ = ground-level concentration of sulfur dioxide in g/m^3 at a point x km
downwind from the stack,

Q = emission rate of SO_2 in g/sec from the top of the stack,

u = wind speed in m/sec,

A and B are constants that depend on the effective height of the smokestack.

The units of measure for $C(x)$ are grams per cubic meter of SO_2. For example, if $C(3) = 13.3$ g/m^3, this means that one cubic meter of air on the ground at 3 km from the smokestack contains 13.3 grams of sulfur dioxide.

For fixed values of Q and u, the effective height of the smokestack determines both the point x_{max} km downwind, where the maximum ground-level concentration occurs, and the maximum concentration C_{max}. We use data taken from curves developed by Turner (Masters 1991) to approximate x_{max} and C_{max} for given effective stack height H. These data are shown in Table 1, and are used to derive the function $C(x)$. This table shows a relationship between effective stack height, the point of maximum ground-level SO_2 concentration, and the maximum concentration for $Q = 1$ g/sec and $u = 1$ m/sec. Here,

H = effective height of the smokestack in meters,

x_{max} = distance in kilometers downwind from the smokestack where the
maximum ground-level concentration occurs, and

C_{max} = maximum SO_2 concentration in g/m^3.

TABLE 1

H (meters)	x_{max} (km)	C_{max} (g/m^3)
20	0.20	350.0
40	0.42	90.0
60	0.70	40.0
80	0.90	24.0
100	1.20	15.0
120	1.40	12.0
140	1.70	7.5
160	2.00	5.5
180	2.30	4.1
200	2.50	3.5
220	2.80	3.0
240	3.00	2.5
260	3.30	2.1
280	3.60	1.7
300	4.00	1.5

These data will be used to derive a function

$$C_1(x) = \frac{Ax}{x^3 + B}$$

which gives the ground level SO_2 concentration when $Q = 1$ and $u = 1$. Then we will only have to multiply or divide (respectively) this function by given values for Q or u to obtain other concentration functions $C(x)$ for different emission rates and wind speeds.

IMPORTANT: The data provided in this table are for $Q = 1$ g/sec and $u = 1$ m/sec.

For a particular stack height, this table can be used to find the distance x_{max} kilometers downwind, where the maximum ground-level concentration of SO_2 occurs and the SO_2 concentration C_{max} at that position. Using these values, it is possible to find the constants A and B for the concentration function,

$$C_1(x) = \frac{Ax}{x^3 + B}$$

when $Q = 1$ g/sec and $u = 1$ m/sec.

Exercise 3. Follow the steps below to determine the ground level concentration function $C(x)$.

(a) Find the values of x_{max} and C_{max} for the effective stack height computed in Exercise 1.

(b) Any function of the form $f(x) = Ax/(x^3 + B)$ (where A and B are positive) has a graph of the shape shown in Figure 3. Using calculus, it can be shown that the maximum point on the graph occurs at $x_{max} = (B/2)^{1/3}$. Therefore, using this equation, if we know where the maximum occurs, i.e., if we know x_{max}, then we can compute B. If we also know the maximum value, i.e., $f(x_{max})$, then we can solve for A. Use this information and your answer to (a) to determine the constants A and B. Then write the concentration function $C_1(x)$ for $Q = 1$ g/sec and $u = 1$ m/sec. Round your answer to 3 decimal places.

(c) The average wind speed in this region is assumed to be 1 m/sec; the emission rate was computed in Exercise 2. Use these values to derive the function

$$C(x) = \frac{Q}{u}\left[\frac{Ax}{x^3 + B}\right].$$

This will be the concentration function we will use throughout the remainder of this module.

(d) Use your calculator to graph the function $C(x)$ and confirm its maximum value.

Not all of the SO_2 emitted from the plant is oxidized to sulfuric acid, and not all of the sulfur emitted is distributed as acid rain. Some remains in the atmosphere and is widely dispersed, and at distances over 15 kilometers from the plant, SO_2 concentrations are insignificant. In the following exercise, we will approximate the percentage of SO_2 distributed over the watershed.

Exercise 4.

(a) Remember that the units of measure for $C(x)$ are grams per cubic meter of SO_2. For example, if $C(3) = 13.3$ grams per cubic meter, a cubic meter of air on the ground 3 kilometers from the smokestack contains 13.3 grams of SO_2.

　　　Assume that the SO_2 *concentration is* $C(x)$ *for each kilometer from* $x - 1$ *to* x. For instance, using the above example, each cubic meter of air from 2 km to 3 km contains 13.3 grams of SO_2.

　　　Now imagine lots of cubic meters of air lined up on the ground from the smokestack to 15 km downwind. Since 1 km $= 10^3$ m, all of the boxes together from 2 km to 3 km contain $13.3 \times 10^3 = 13{,}300$ grams of SO_2. We are concerned primarily with the 15 km $\times$ 1 m $\times$ 1 m "long box" directly downwind from the stack and the portion in the 3 km $\times$ 1 m $\times$ 1 m "short box" over the watershed (see Figure 4.)

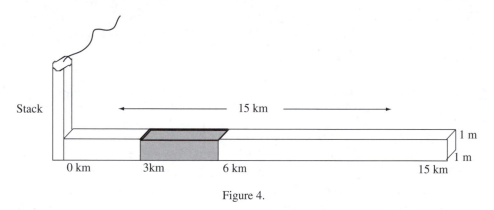

Figure 4.

　　　We assume that the percentage of SO_2 distributed over the total watershed is the same as the corresponding percentage directly downwind, that is, the ratio

$$\frac{\text{amount of } SO_2 \text{ distributed over the watershed directly downwind}}{\text{total amount of } SO_2 \text{ distributed directly downwind}} \times 100.$$

We will ignore any SO_2 deposited more than 15 kilometers downwind.

　(i) Approximate the amount of SO_2 in the "long box" of air.

　(ii) Approximate the amount of SO_2 in the "short box" of air over the watershed.

(b) Use your answers from (a) to approximate the percentage of SO_2 that lies over the watershed. (Although this approximation only takes into account the sulfur dioxide directly downwind from the power plant, we assume that the same ratio holds over the entire region.)

2　ACID RAIN

You are now ready to begin work on the determination of the overall effect of SO_2 emission on Lonesome Lake. The next step is to figure the acidity of precipitation in the region.

The acidity of a liquid is measured by its hydrogen ion concentration, and is denoted by pH. If z denotes the hydrogen ion concentration in moles per liter, then

$$pH = -\log z,$$

where log refers to common logarithm. The pH varies from 0 to 14; a graph is shown in Figure 5.

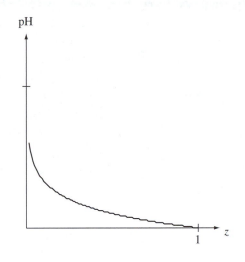

Figure 5.

Note that as the hydrogen ion concentration z increases, the pH decreases, and since z is between 0 and 1, $pH = -\log(z)$ is positive. Since the larger the hydrogen ion concentration the more acidic the liquid, then small values of pH indicate an acid liquid, whereas large pH values indicate alkalinity. Water with $pH = 7$ is neutral and considered "pure," whereas water with pH smaller than 7 is considered acidic. Thus for pure water, the hydrogen ion concentration is 10^{-7}, and

$$pH = -\log(10^{-7}) = -(-7) = 7.$$

The steps below will lead you to the determination of the pH of precipitation over the watershed. For simplicity, the only sulfur dioxide source we consider in these calculations is the emission from the power plant. We assume:

1. the total emission of SO_2 from the plant is $10^{7.81}$ grams per year;
2. the total amount of water from the precipitation over the watershed is $10^{9.45}$ liters per year; and
3. one-fourth of the airborne sulfur is deposited in precipitation. (The rest, which is either deposited gradually as precipitation or dispersed in the atmosphere, will not be considered in this study.)

Exercise 5.

(a) The total annual emission of SO_2 from the plant is $10^{7.81}$ grams per year. Our primary concern is the amount of sulfur emitted, since sulfur is the component that forms the sulfuric acid. The atomic weight of one sulfur atom is 32.07 awu (atomic weight units), whereas one oxygen atom has atomic weight 16.00 awu, so sulfur contributes approximately one-half the weight of sulfur dioxide. Find the annual emission of sulfur from the plant (measured in grams).

(b) Since pH measures hydrogen ion concentration in moles per liter, the sulfur must be measured in *moles*. Determine the number of moles of sulfur in the annual SO_2 emission from the power plant. You need this information: one mole of a substance is equal to its molecular weight in grams; so 32.066 grams of sulfur equal one mole.

(c) Recall that the total amount of water from precipitation over the watershed is $10^{9.45}$ liters per year and approximately one-fourth of the airborne sulfur is deposited in precipitation. Use your answers to (b) above and (b) in Exercise 4 to compute the concentration of sulfur in moles per liter in precipitation over the watershed.

(d) For every molecule of sulfuric acid, two hydrogen ions are produced; i.e. for each molecule of H_2SO_4 (which contains one sulfur molecule), two hydrogen ions (H^+) result. Determine the concentration (in moles per liter) of hydrogen ions in precipitation over the watershed.

(e) What is the pH of the precipitation?

3 ACIDITY OF THE WATER IN LONESOME LAKE

Two things affect the pH of the precipitation before it reaches Lonesome Lake. Some of the precipitation immediately evaporates, and also the surface water in the lake evaporates. Both of these further concentrate the sulfuric acid. Overall, a total of one-fourth of the total precipitation evaporates. In this process the amount of sulfuric acid remains constant so that there is an even higher concentration in the lake water. Thus the concentration of H+ ions is higher and the pH is lower. In the steps below, you will determine the acidity of the water in the lake; we assume that no other chemical reactions take place that further affect the pH of the water.

Exercise 6.

(a) Determine the hydrogen ion concentration of the water after the evaporation.

(b) What is the pH of the water in Lonesome Lake?

4 FISH

It has been established that acid water is harmful to fish life. In particular, a study of declining fish populations in acidified freshwater lakes in Norway was conducted by Leivestad, Hendrey, Muniz, and Snekvik in the 1970s. The findings were conclusive; fish simply do not survive in waters that have low pH. The disappearance of fish in freshwater is not only ecologically unsound, but also affects commercial and recreational activities.

Exercise 7. Table 2 indicates some of the results of the Leivestad et al. study. Examine this table and determine the probability that Lonesome Lake has any fish. Then examine the variables involved in this module to see what can be done to improve this probability. The variables are:

(i) emission rate Q (this could probably be reduced some, but if you want electricity, not too much);

(ii) wind speed (of course you have no control over this, but you can see how increased wind speed affects dispersion of sulfur dioxide);

(iii) stack height H (this is one of the main factors used in controlling SO_2 dispersion from coal burning power plants).

Write a report explaining your plan to reduce the acidity of Lonesome Lake so that the fish have a chance.

TABLE 2

pH	% Lakes with no fish
< 4.5	75
4.5–4.7	55
4.7–5	35
5–5.5	30
5.5–6	15
> 6	5

REFERENCES

Harte, John, 1988. *Consider a Spherical Cow, A Course in Environmental Problem Solving*. Mill Valley, CA: University Science Books.

Park, Chris C., 1987. *Acid Rain: Rhetoric and Reality*. London and New York: Methuen. p. 85.

Masters, Gilbert, 1991. *Introduction to Environmental Engineering and Science*. Englewood Cliffs, New Jersey: Prentice Hall. pp. 317–324.

Using Math in Environmental Detective Work

Charles R. Hadlock
Bentley College, MA

INTRODUCTION

Today, with the Earth's environment in questionable health, mathematics is playing a key role in solving all kinds of environmental problems. One of the most important uses of mathematics is in groundwater hydrology, the study of how water moves where you can't see it—beneath the surface of the Earth.

You probably already know that if you started digging a hole almost anywhere—in your backyard, out in the school yard, in the middle of a big city—you would eventually hit water. It might happen just a few feet below the surface, or it might take hundreds of feet of digging; maybe you would even have to break through solid rock. But yes, you would eventually hit what's commonly called groundwater filling all the little cracks or spaces between the particles.

Now, that groundwater isn't just sitting there! It's actually moving. It's not rushing along like a river on the surface; in fact, it might just creep along at only a few inches a day. But fast or slow, it's moving—so if it ever gets contaminated by anything that leaks into it, that contamination is off and running, too. Since much of our drinking water comes from wells that draw on groundwater, this contamination could eventually become a real problem.

As you work through this unit, your job is going to be to figure out where the groundwater's headed and how fast it's going. Now that's going to be quite a trick ... because you can't actually see it, way down there below the Earth's surface. And that's where mathematics comes to the rescue!

In the activities that follow, you'll be using mathematical formulas to calculate a variety of hydrologic values as you investigate a hypothetical gasoline leak and the potential groundwater contamination it represents. You'll be challenged to determine the direction and flow of groundwater in the area, and track the potential spread of contaminants into the water supply—just as a groundwater hydrologist would.

MISSION PROFILE

You will be working on an imaginary mission in this unit that will take you on a visit to the property of the fictitious Mr. Schmarmer, a farmer in Couldbeany County, USA. County officials have called you in to investigate a leak from an underground gasoline storage tank at a service station near Farmer Schmarmer's land. Using the data and mathematical formulas provided in the course of your investigation, you'll learn to determine, among other things:

1. The **flow** direction of the area's groundwater—and possibly of the leaking gasoline.
2. The **velocity** or speed at which the groundwater (and possibly the gasoline) is traveling.
3. **If the gasoline will contaminate** any wells on Farmer Schmarmer's property.
4. **How long** it might take the gasoline to get to those wells if nothing is done to stop it.

There's your mission. To complete it, work through each of the following sections using the information and examples provided. So get to it . . . and good luck!

GROUNDWATER BASICS

Porosity

Let's start with a few basic terms. Among the most important is **porosity**, the ability of a soil or rock material to store water. Porosity is measured as the ratio of spaces or openings in the soil or rock to the total volume of material. It largely depends on the size and shape of the soil or rock particles.

Porosity is usually expressed as either a decimal fraction or as a percentage. So if someone tells you that a bucketful of sand has a porosity of .4 or 40 percent, that means that 40 percent of the volume is really air spaces between the grains. Underground, this 40 percent could be filled with water.

The Water Table

Except in a really wet swampy area, you have to dig down into the ground a little bit before you reach the zone where the pores are completely full of water. There will always be at least a little moisture in the soil as you dig down, but when you first reach a zone where water actually starts to seep into the hole you dig, you have reached the **saturated zone**, and the top border of it is called the **water table**.

Aquifers

If you keep going down, you might occasionally dig or drill through some layers of soil or rock that are nearly impervious to water, but after passing through them, you might again encounter layers where water would again seep into your hole. All the layers at which water seeps in would be called **aquifers**, and mathematics is used all the time to figure out how the water is moving within them. For your mission, we will concentrate on the top aquifer, which is generally called an **unconfined** or **water-table aquifer**.

KEY HYDROLOGIC VALUES

The depth of the water table in an unconfined aquifer is an important piece of information for hydrologists, since it directly affects both the use of land and the development of water supplies. A shallow (i.e., high) water table can result in the land becoming "waterlogged" and unusable for many purposes, while a very deep (i.e., low) water table can make it expensive and impractical to drill and use a water well.

To determine the height of the water table, the water level in a well is measured from the land surface, and this distance is expressed in feet or meters. This figure is then used by the hydrologist (in this case, you) to determine a number of key hydrologic values that will come in handy for your mission.

Hydraulic Head

The first of these values is known as the **hydraulic head** of the well. For an unconfined aquifer, this is simply the height or elevation of the water table at any point, measured with respect to a common level (such as sea level). The hydraulic head is easy to calculate if you know the elevation of your measuring point (usually the land surface) and how far the water table is beneath that measuring point. For example in Figure 1, calculate the head h_1 of Well 1 as follows:

$$h_1 = 100 \text{ ft} - 16 \text{ ft} = 84 \text{ ft}$$

because **100** is the elevation of the measuring point (in feet)

and **16** is the depth to water in feet below the measuring point.

Similarly, you'd determine the head h_2 for Well 2 as follows:

$$h_2 = 96 \text{ ft} - 18 \text{ ft} = 78 \text{ ft}$$

Hydraulic Gradient

Once you have calculated the hydraulic head for several wells, you are ready to determine another key factor: the **hydraulic gradient**, or the water table's slope. The hydraulic gradient is really the same as the **slope** of a straight line, except that we always refer to it as a positive number and state the direction in which it's going down.

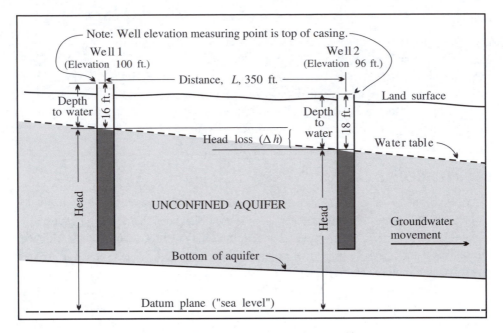

Figure 1. Cross-section of unconfined aquifer.

Knowing the hydraulic gradient is important because water flows in an unconfined aquifer just like a ball rolls down a hill: **the steeper the slope, the faster it wants to go**. In fact, if you think about letting go of a ball on the side of a hill, you also know that it will start rolling down in the steepest direction from any point. Water in an aquifer wants to do the same thing—**it wants to flow in the direction in which the hydraulic head decreases the fastest**. So this is the direction in which we will want to calculate the hydraulic gradient. If it's large (or steep), then that means the water would be trying to move faster there.

Let's go back to Figure 1 and calculate the hydraulic gradient for the two wells pictured there. The gradient is obtained by dividing the head loss, or the difference in head between the two wells, by the distance between them. This can be expressed in the following equation:

$$i = \frac{\triangle h}{L} = \frac{(h_1 - h_2)}{L}$$

where

i = the hydraulic gradient

$\triangle h$ = the head loss between two wells

L = the distance between the two wells (in ft)

h_1 = the head for Well 1

h_2 = the head for Well 2

Plugging in the numbers from Figure 1, we can compute this value:

$$i = \frac{\Delta h}{L} = \frac{(84 - 78)}{350} = \frac{6}{350} = .017$$

So, expressed as a decimal, the hydraulic gradient for these two wells is **.017** from Well 1 to Well 2.

Exercise 1. *Determining Hydraulic Gradient*   Now, take your first step toward solving the hydrologic "Mystery of Schmarmer's Farm": calculate the **hydraulic gradient** between two test wells you've sunk into the unconfined aquifer under Farmer Schmarmer's property, as shown in Figure 2. (You can assume that groundwater moves from Well A to Well B). Use the process we've just reviewed to determine the head for each well, and then calculate the hydraulic gradient.

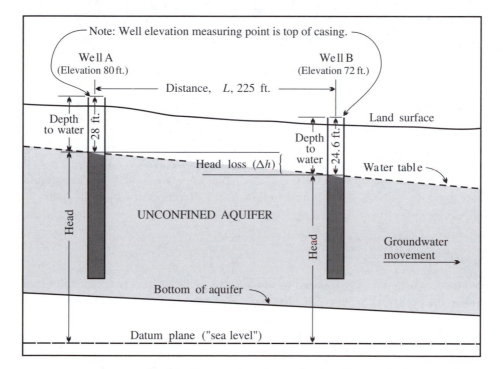

Figure 2. Cross-section of unconfined aquifer.

GROUNDWATER FLOW DIRECTION
Contour Maps

Hydrologists usually sink a number of test wells into an aquifer, calculate their head values, then put this information together to form a **water table contour map** like the one in Figure 3. Incidentally, if you've ever used a topographic map to go on a hike, it's really the same thing. On a "topo" map, the lines show points of equal elevation; on a water table contour map, the lines show points where the head values are the same, and their values are labeled on the map.

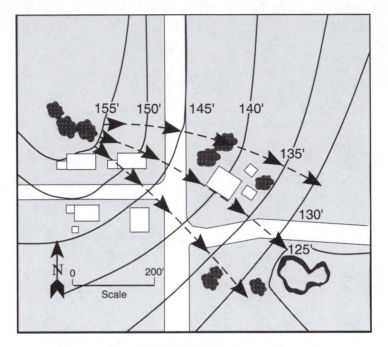

Figure 3. Water table contour map.

Flow Direction

To determine flow direction with a contour map, you can draw lines **perpendicular** to the contour lines at several points. The direction in which the head value decreases the fastest from any point is always the direction perpendicular to the contour line. This is the direction the groundwater will move in (except in some very peculiar geologic environments outside the scope of this introduction).

Figure 3 shows how this works. The solid lines are contour lines for the hydraulic head. The broken lines are flow lines for groundwater. At every point where they intersect, they must be perpendicular!

A Simple Way to Estimate Flow Direction

Drilling wells can be pretty expensive, so it's very common to try to get by with as few as you can. In fact, even if you only have three wells, there's a clever way to get a rough idea of which way the groundwater's going. All you need is the following information:

- The wells' relative geographic positions.
- The distances between them.
- The head value of each well.

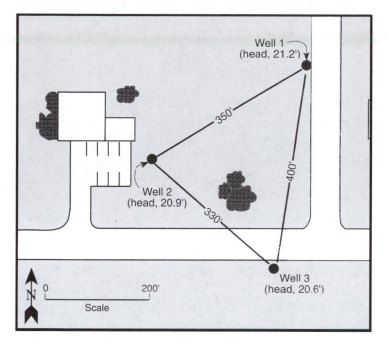

Figure 4. Site map with head values for three wells.

We have all these facts for the three wells shown in Figure 4. Now we can use these facts to calculate where a contour line would lie… and that will enable us to determine flow direction.

Let's start our calculations following the logic outlined below. The steps we'll be taking are illustrated in Figure 5.

(a) First, you determine which well has the intermediate (or middle) head value of the three; in this case, it's Well 2, with a head of 20.9.

(b) Now you make a key observation: **Somewhere along the line from Well 1 to Well 3, there must be another point where the head is the same as it is at Well 2.** This is because the head must gradually change from a higher value at Well 1 to a lower value at Well 3. So it must hit the value 20.9 somewhere in between.

You can estimate this exact point by using the simple principle of proportions. Look at it this way: if you were looking for the point where the head value was halfway between the Well 1 value (21.2) and the Well 3 value (20.6), you'd expect it to be halfway along the 400-foot distance between the two wells. (If you were looking for the point where the head value was two-thirds of the way between the Well 1 value and the Well 3 value, you'd go two-thirds of the distance, or (2/3)400 = 267 feet down from Well 1 to Well 3.)

Applying this principle to the wells in our diagram, ask yourself where along the range from 21.2 (Well 1) to 20.6 (Well 3) is the head value of Well 2 (20.9)? The answer: it's exactly *halfway* along, because:

$$\frac{\text{desired difference}}{\text{total difference}} = \frac{22.2 - 20.9}{21.2 - 20.6} = \frac{.3}{.6} = \frac{1}{2}$$

Therefore, we should go half the 400-foot distance to get to the point we are looking for. This point 200 feet from Well 1 is where we estimate the hydraulic head is 20.9, the same value as at Well 2.

(c) Now we can use this point to indicate a contour line, by connecting it to Well 2 with a straight line. This line represents the **approximate** water-level contour where the head is always equal to 20.9, the head at Well 2.

(d) Since groundwater will move in a direction **perpendicular to** the contour line you've just established, you can now find the general groundwater flow direction by drawing lines or arrows perpendicular (90°) to the water-level contour. In Figure 5, you'll see that we've drawn three such arrows, one of them directly through Well 3.

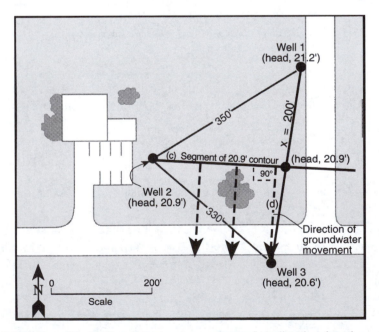

Figure 5. Calculation of hydraulic gradient and groundwater flow direction, based on map in Figure 4.

How long do you think this last arrow is? Well, since it's a perpendicular line, it must be shorter than 200 ft, the distance along the original line. You can find the exact value if you know trigonometry, but we'll just use the scale on the figure, which still makes it look very close to 200 ft, a good enough approximation.

Finding the Gradient at the Same Time

While you're calculating the flow direction, you can also determine the hydraulic gradient by dividing the head difference between the well with the lowest head (Well 3, 20.6) and that of the contour (20.9) by the approximate distance between them (200 ft). This is why we needed to estimate the length of that third arrow just above. This will yield the hydraulic gradient, as follows:

$$i = \frac{\triangle h}{L} = \frac{20.9 - 20.6}{200} = \frac{.3}{200} = 1.5 \times 10^{-3} \text{ ft per ft}$$

Now that you see how valuable that approximate distance of 200 was to getting the hydraulic gradient, it would be a good exercise to try to find it exactly if you have studied trigonometry.

Exercise 2. *Determining Flow Direction and Hydraulic Gradient on a Site Map* Now take a look at Figure 6, which represents a site map with water table elevations (heads) for three wells in an area you're investigating on Farmer Schmarmer's property. Here's your next challenge: develop a contour line, then determine the groundwater flow direction, and finally calculate the hydraulic gradient in the flow direction. Use the steps and equations in

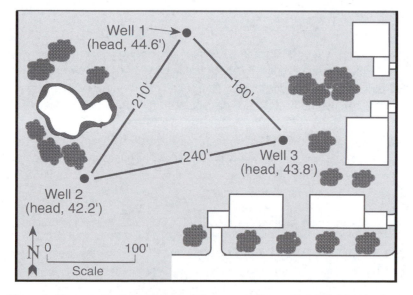

Figure 6. Site map for three wells on Schmarmer's Farm with labeled head values.

the example we just completed. After developing and drawing in your contour line, draw arrows and lines to represent the direction in which the groundwater would flow. Then calculate the hydraulic gradient along these **flow lines**. (Use a ruler when applying the map scale, or try out your trigonometry again if you're good at it.)

GROUNDWATER FLOW RATES

Hydraulic Conductivity

Now that you've seen how hydraulic gradient and flow direction figure into your hydrologic investigation of Schmarmer's Farm, you're ready to learn about **hydraulic conductivity**—a measure of water's ability to move through a certain type of soil or rock. As the result of many experiments conducted by hydrologists, typical ranges of values for different kinds of underground formations have been developed. See the table below for some such values, as well as accompanying values of porosity.

Geologic medium	Hydraulic conductivity (K, ft/day)	Porosity (%)
Gravel	100–100,000	25–40
Sand	0.01–100	25–50
Silt	0.0001–0.1	35–50
Clay	0.0000001–0.001	40–70
Sandstone	0.00001–0.1	5–30
Limestone	0.0001–0.1	0.01–20
Granite (weathered or fractured)	0.001–10	0.01–10

Caution: Even though the units of hydraulic conductivity K are feet per day, which makes it sound like a speed or velocity, the K value itself does not represent the speed at which the groundwater is moving. After all, you would expect some additional information to enter into the calculation of the velocity, especially the hydraulic gradient, which was discussed earlier.

But hydraulic conductivity does represent how easily a given material can permit water to flow through it, and so it certainly is used as part of the calculation of groundwater velocity. It is also used to determine groundwater flow rates, using an equation known as **Darcy's Law**, which is actually our next topic.

Darcy's Law

In 1856, French engineer Henry Darcy first expressed the factors governing movement of groundwater mathematically, in the following equation now referred to as Darcy's Law:

$$Q = K A i$$

where

Q = the quantity of water per unit of time flowing through the part of the aquifer under study;

K = the hydraulic conductivity;

A = the cross-sectional area through which water is transmitted (as shown in Figure 7);

i = the hydraulic gradient.

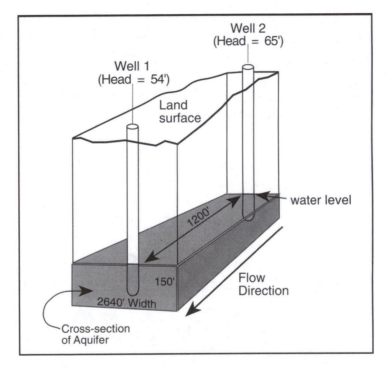

Figure 7. Cross-section of aquifer showing width and well placement.

Just look at the simple logic behind this equation. Do you remember from algebra that a mathematical way to say that one thing is **proportional to** another is to write the first as a constant times the second? This is really what Darcy's Law does. Think of K as the constant of proportionality. Then the equation says that the total amount of flow is proportional to two things: the total cross-sectional area of the part of the aquifer you are interested in, and the hydraulic gradient (that is, the steepness of the slope of the water table). This really makes sense. For example, if you want to measure the total amount of water flowing in twice as big a section of the aquifer, you would expect to find twice as much. Or, if you double the steepness of the water table, so that there is more force making the water move, then you would also double the amount of flow.

Calculating Flow Rate

Consider the example in Figure 7: You have an aquifer that is 150 ft thick, and one-half-mile (2,640 ft) wide. Two observation wells are located 1,200 feet apart in the direction of groundwater flow. The two hydraulic heads are 65 ft and 54 ft. The hydraulic conductivity is 80 ft/day. What is the total daily flow of water (Q) through the aquifer? Using Darcy's Law, you'd solve the problem like this:

$$Q = K \times A \times i$$

$$= 80 \text{ ft/day} \times (150 \text{ ft} \times 2640 \text{ ft}) \times \frac{(65 - 54)}{1200}$$

$$Q = 290{,}400 \text{ ft}^3/\text{day}$$

Exercise 3. *Determining Flow Rates with Darcy's Law*   Darcy's Law is one of the most important equations in your hydrologist's "bag of tricks." Here's a chance to put it and the other information you've gathered to good use. Our friend Farmer Schmarmer has just informed you that he's noted a strange taste in his drinking water! So you're going to study the aquifer under his property.

Use the information in Figure 8. As you can see, the coarse sand aquifer below the farm is 100 ft thick and 3/4 mile (3,960 ft) wide. You've sunk an observation well 1,000 ft

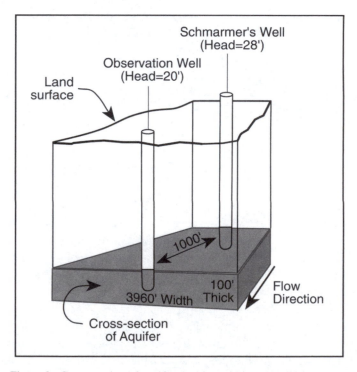

Figure 8. Cross-section of aquifer showing width and well placement.

away from the farmer's well, in the direction of groundwater flow. The hydraulic head for the observation well is 20 ft; Schmarmer's well has a head of 28 ft. The hydraulic conductivity of a coarse sand aquifer is about 100 ft per day, the high end for sand as shown in the table. So the question is: what is the total daily flow of water (Q) through the aquifer? Use Darcy's Law to find out!

GROUNDWATER VELOCITY AND TRAVEL TIME

Now we're ready to fit the last piece into the puzzle by calculating **groundwater velocity**, the speed at which groundwater travels through an aquifer. As you might imagine, finding groundwater velocity is key to determining how long it would take groundwater—which might be carrying a contaminant—to flow from one point to another.

Calculating Velocity

Groundwater velocity is obtained through the following simple equation, adapted from Darcy's Law, known as the **interstitial velocity equation**:

$$v = \frac{Ki}{\eta}$$

where

v = the velocity of the water in the flow direction

K = the hydraulic conductivity;

i = the hydraulic gradient; and

η = (the Greek letter eta, pronounced "a'-ta") is the *porosity* of the material (the ratio of openings to the total volume of the soil or rock)

Here's an example of how this equation can be used to determine the velocity of water through a typical good (that is, carrying lots of water) aquifer.

K (hydraulic conductivity) = 30 ft/day

i (hydraulic gradient) = 0.006

η (porosity) = 0.20

$$v = \frac{Ki}{\eta} = \frac{(30 \text{ ft/day})(0.006)}{0.20} = .9 \text{ ft/day}$$

So the groundwater velocity is 0.9 ft/day. And once you have the velocity, it's easy to figure out how long it would take the water to cover a certain distance. Just think about driving a car. If your velocity is 60 miles per hour and you have to go 90 miles, the time it takes would be:

$$\text{time} = \frac{\text{distance}}{\text{velocity}} = \frac{90 \text{ miles}}{60 \text{ miles per hour}} = 1.5 \text{ hours}$$

In our groundwater example, we found that velocity was .9 ft/day. Therefore, if we wanted to know how long it would take water to go, say, 500 feet, we would calculate as follows:

$$\text{time} = \frac{\text{distance}}{\text{velocity}} = \frac{500 \text{ feet}}{.9 \text{ ft/day}} = 556 \text{ days}$$

So you can see that it might take a year or more for groundwater to travel a relatively short distance. That's why a lot of groundwater contamination problems go unnoticed for a long time after the pollution starts. Unfortunately, when they finally are discovered it can be terribly difficult and expensive to do anything about them.

Exercise 4. *THE FINAL CHALLENGE: Determining Groundwater Travel Time*   Now, to completely solve the Schmarmer's Farm case, you'll need to estimate the time the groundwater contaminated by gasoline from a leaking tank at the gas station—over a mile away—will take to reach one of Farmer Schmarmer's wells, based on groundwater velocity and other values you worked with earlier.

The contour map in Figure 9 shows estimated contours of hydraulic head in a shallow sand aquifer below Schmarmer's farm (porosity = .25, with a hydraulic conductivity of 102 ft/day). Suppose that the gasoline tank leak occurred at point X on the map. You are to determine:

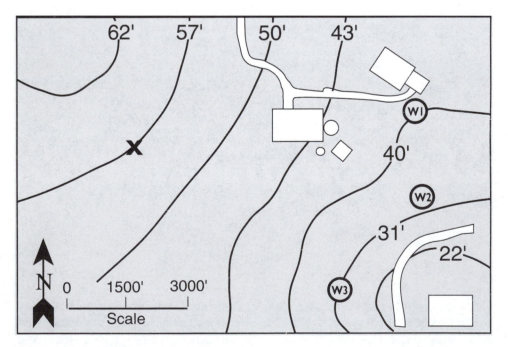

Figure 9. Water table contour map, showing source of contaminant and Schmarmer's three wells.

(a) The path contaminated groundwater would take.

(b) Which of Schmarmer's wells is in danger (W1, W2 or W3).

(c) How long it might take the contaminated groundwater to reach the well.

Use the diagram to work out your answers. And remember these tips: use the contour lines and the head value for the wells and the gas tank to calculate the *flow path* from Point X. Then break up the flow path into segments from each contour line to the next. For each of these segments, calculate the distance (using the scale), the hydraulic gradient, interstitial velocity and travel time. Then combine your results to answer the third part of the challenge.

When you've figured this one out, you've completed your mission. Congratulations!

FIELD ACTIVITIES

If you want to see mathematics in action along the lines discussed above, you can probably find people working on real problems in your home community similar to the fictitious situation of Schmarmer's farm. To track this down, call your local Board of Health or any office of your state's environmental agency. Ask them if there are any state or federal "Superfund" sites near you. (There are thousands, perhaps tens of thousands in the country, so you probably have some nearby.) These are sites that have been contaminated underground by leaking chemicals, buried wastes, or similar things. Since most of these sites are being overseen by government agencies, you will find that it's often easy to arrange a visit and see first hand how interdisciplinary teams of modelers, geologists, engineers, lawyers, doctors, economists, politicians and others make decisions about what to do about such sites.

Another interesting activity is to find a videotape on groundwater pollution problems. One good example is a PBS *NOVA* program on the very complex contamination situation in Woburn, Massachusetts. The name of this videotape is "Toxic Trials," and it may even be in your school or town library. But there are also many others. Your school librarian or media specialist should be able to help.

Government agencies are usually eager to educate the public about the steps being taken to improve the environment, and mathematical modeling is one of the tools that is used in the analysis of almost every kind of environmental problem. If you contact such an agency, it is likely that they would provide a speaker or arrange a tour for you to learn more about these situations. Such agencies include the EPA, Army Corps of Engineers, the Coast Guard, as well as state environmental agencies.

The unifying theme you should expect to see in any of these "real world" situations is that they are interdisciplinary. People have to learn about many different fields, and they have to be able to communicate in ways that can be understood by people in different fields. That's why it is very valuable when you are in school to take courses in a wide variety of subjects. This will provide a good basis to let you make valuable contributions to the solutions of these complex, interdisciplinary situations in the future.

Modeling Ground-Level Ozone with Basic Algebra

James V. Rauff
Millikin University, IL

1 INTRODUCTION

In this module you will learn how to construct, use, and evaluate a linear model of trends in the ground-level ozone in Southern California between 1992 and 1995.

Mathematical models describe relationships between observed or measured quantities. Models are used to predict the weather, to guide prudent investment decisions, to formulate and test public policy, to identify societal trends, to help manage forests and parks, to design automobiles, and to aid in a wide variety of other useful scientific and business activities.

Ozone (O_3) is a form of oxygen that consists of three oxygen atoms. Ozone in the upper atmosphere protects life on earth by filtering out harmful ultraviolet radiation from the sun, but at ground level, ozone is a pollutant. Ground-level ozone is formed through a complex of chemical reactions between hydrocarbons, nitrogen oxide, and sunlight. These hydrocarbons and nitrogen oxides come from industrial processes and the engine emissions of cars, buses, trucks, construction vehicles and boats. Ozone is the major component of smog.

When inhaled, ozone can damage the lungs. It can cause chest pains, coughing, and nausea. Ozone can also worsen bronchitis, emphysema, and asthma. Ground-level ozone also affects plant life. Ozone reduces the ability of trees and other plants to fight disease; it interferes with the production and storage of starches within plants, stunting their growth, and it damages the quality and yield of commercial crops like corn, wheat, and soybeans.

55

Because of the public health risks of ground-level ozone, the U.S. government and many states have set health standards for the amount of ozone in the atmosphere and have instituted monitoring programs to see if the standards are being met. The Clean Air Act of 1990 requires the U.S. Environmental Protection Agency, states, and cities to devise and implement programs that will reduce hydrocarbon and nitrogen oxide emissions from cars, fuels, industrial facilities, power plants, and other sources.

2   READING REAL DATA: OZONE DATA FROM AQMD

The South Coast Air Quality Management District (AQMD) is the primary smog control agency for the 15 million people who live in Los Angeles, Orange, Riverside, and San Bernadino counties in California. Table 1 is taken from the AQMD's table of "Historic Ozone Air Quality Trends," which can be found at the AQMD website http://www.aqmd.gov. The abbreviation "ppm" means "parts per million." This is a way of measuring how much ozone is actually in the air. Think of a large container containing 1,000,000 gallons of clean air. This container would be about the size of a box 6 feet high whose base was about half the size of an American football field. Now take a pint (1 pint = .125 gallons) of this clean air out and replace it with a pint of ozone and mix thoroughly. The large bottle now has a ozone concentration of .125 ppm.

The U.S. government believes that any ozone concentration over .12 ppm is dangerous to one's health, so the Federal Standard is .12 ppm. California has a stronger standard of .09 ppm. A Health Advisory warns people with respiratory problems to stay indoors, and people planning athletic activities to be especially careful because the ozone level is high.

TABLE 1   Ozone Trends 1989–1995. Number of Basin Days* Exceeding Health Standard Levels

Year	State Standard (0.09 ppm)	Federal Standard (.12 ppm)	Health Advisory (.20 ppm)
1989	211	157	120
1990	184	130	107
1991	183	130	100
1992	191	143	109
1993	185	124	92
1994	165	118	96
1995	154	98	59

*Basin days represent the number of days when a standard was exceeded anywhere in the South Coast Air Basin.

EXERCISES

2.1. For how many days was the State Standard exceeded in 1990? In 1994?

2.2. Considering only ozone, was the overall air quality better or worse in 1994 than it was in 1990?

2.3. Look at the whole period from 1989 to 1995. What can you say about the ozone air quality over this period?

3   LINEAR RELATIONSHIPS: OZONE AIR QUALITY TRENDS

When scientists and public policy makers want to get a good understanding of trends, they will often make a graph of the data they have collected. Long term trends can often be very complicated, so we often use smaller intervals to get "snapshots" of what is happening. In this case of ozone air quality we can see that the number of days that the State Standard was exceeded has been decreasing since 1992. We may wish to ask what we can predict for later years if this decrease continues.

Figure 1 is a plot of that portion of the State Standard data from Table 1. The graph shows the number of days that the State Standard was exceeded for the years 1992–1995. In order make the graph easier to read, I have used the common technique of designating one year as year 0 and then numbering each year after that accordingly. Since 1989 is the first year in our data, I'll call 1989, year 0. Then 1990 is year 1, 1991 is year 2, 1992 is year 3, 1993 is year 4, 1994 is year 5, and 1995 is year 6.

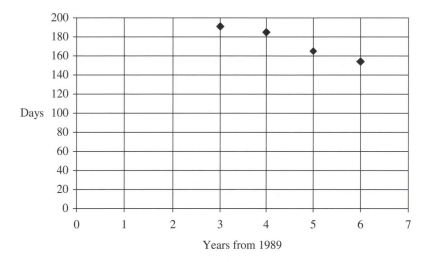

Figure 1. State Standard Data 1992–1995

You can see that the number of days in which the State Standard was exceeded have been generally declining since 1992. You can also see that the points on the graph are approximately lined up and that the points on the graph are getting lower as we read the points left to right.

EXERCISE

3.1. If the trend in ozone air quality from 1992 to 1995 continued for 1996, which of the following would be the best reasonable estimate for the number of days in which the State Standard was exceeded in 1996? Why?

    (a) 195

    (b) 110

    (c) 145

    If the points in Graph 1 were perfectly lined up, then we would say that there is a **linear** relationship between the year and the number of days when the State Standard was exceeded. If the points are not perfectly lined up, then we can produce only an approximate model. An approximate model will have a graph that comes close to most of the actual data points. We can find such a model by sketching and by using some algebra. A sketch model attempts to draw a graph of a simple curve close to all the data points. In this module we are concerned only with linear models, so our sketches will all be straight lines.

    In Figure 2, you can see a straight line, called Model 1, drawn through the points representing the data for the years 1992 and 1993. Notice that this line misses the other two points.

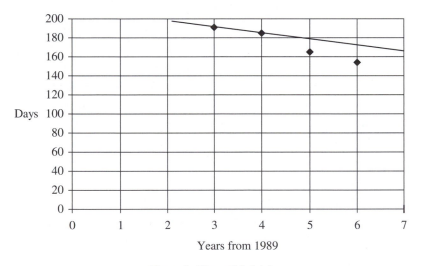

Figure 2. Linear Model 1

EXERCISES

3.2. Find the point where Model 1 crosses the vertical line representing the year 1994 (year 5). Read across from this point to find the number of days that the State Standard was exceeded in 1994 as predicted by Model 1. Is this more or less than the actual number of days?

3.3. Draw a straight line connecting the two points representing the data for the years 1992 and 1994 in Graph 1. Make sure to extend the line beyond the two points. Call this line Model 2. How did Model 2 compare to Model 1 when you consider how close it came to the actual data points?

3.4. Find the point where Model 2 crosses the vertical line representing the year 1993 (year 4). Read across from this point to find the number of days that the State Standard was exceeded in 1993 as predicted by Model 2. Is this more or less than the actual number of days?

3.5. Which of the two lines, Model 1 or Model 2, seems to be a better representation of the actual trend in ozone air quality? Why?

3.6. What does each model predict for the number of days that the State Standard was exceeded in 1996?

We can often get an idea of how good a model is by examining and testing the graph as you did in the previous exercises. A good model will come close to or hit the actual data points. However, you probably had some difficulty in reading the precise number of days that the State Standard was exceeded from the graph. In addition, the concept of "close to the actual data points" is vague. Models that are expressed algebraically rather that graphically often can erase a lot of this difficulty and vagueness.

4   CONSTRUCTING A LINEAR MODEL

Our next task is to come up with a mathematical description of the two linear models we drew in the last section. Recall that a linear relationship between two quantities x and y can be expressed mathematically as an equation that has the form

$$y = \mathrm{m}x + b \tag{1}$$

A linear model of the number of days when the State Standard was exceeded as a function of the years since 1989 would be an equation like (1), where the variable x represents *the year* and variable y represents *the number of days when the State Standard was exceeded in year x*. So, for example, Model 1 would have an equation that would produce the following table:

x: Years since 1989	y: Number of days when the State Standard was exceeded in year x
3	191
4	185

EXERCISE

4.1. Complete the table below for Model 2.

x: Years since 1989	y: Number of days when the State Standard was exceeded in year x

Let's review what is necessary to produce a linear equation. We need two points that line on the graph of the linear equation because at least two points are needed to draw a straight line. The two points will give us the slope of the line and from there we can work our way to the equation. Here's the process of getting the equation for Model 1.

Step 1. Identify the two points.

For Model 1 the two points are (3, 191) and (4, 185). **Remember** to put the x value first and the y value second in each pair.

Step 2. Calculate the slope.

The slope is calculated by taking the difference in the y-values divided by the difference in the x-values. For Model 1

$$\text{slope} = m = \frac{185 - 191}{4 - 3} = \frac{-6}{1} = -6$$

The sign (positive or negative) of the slope indicates the vertical direction of the line as we move left to right. A line with a positive slope tilts up and a line with a negative slope tilts down. *Always make sure that the sign on the slope is correct before going to the next step!* A negative slope makes sense for our Model 1 because the line we sketched in Figure 2 is tilted downward.

Step 3. Substitute the slope and the coordinates of one of your points into the equation $y = mx + b$ and solve for b.

We'll use the slope -6 and the 1992 point (3,191):

$$y = mx + b \qquad \text{(Equation 1)}$$
$$191 = -6 \cdot 3 + b \qquad \text{(Substitution)}$$
$$191 = -18 + b \qquad (-6 \cdot 3 = -18)$$
$$191 + 18 = b \qquad \text{(add 18 to both sides)}$$
$$209 = b \qquad (191 + 18 = b)$$

Thus, the equation for Model 1 is $y = -6x + 209$.

EXERCISES

4.2. Construct the equation for Model 2.

4.3. Use the points representing the data for 1992 and 1995 to construct a third model for ozone air quality trends. Call this Model 3.

4.4. Use the points representing the data for 1993 and 1994 to construct a fourth model for ozone air quality trends. Call this Model 4.

4.5. Use the points representing the data for 1993 and 1995 to construct a fifth model for ozone air quality trends. Call this Model 5.

4.6. Use the points representing the data for 1994 and 1995 to construct a sixth model for ozone air quality trends. Call this Model 6.

4.7. Complete the following table.

TABLE 2   Models of Ozone Air Quality Trends 1992–1995

Model Name	Years Used to Construct Model	Equation of the Model
Model 1	1992 & 1993	$y = -6x + 209$
Model 2	1992 & 1994	$y = -13x + 230$
Model 3	1992 & 1995	
Model 4	1993 & 1994	
Model 5	1993 & 1995	
Model 6	1994 & 1995	

5   INTERPRETING MODELS: TESTING

We now have six different linear models for ozone air quality trends between 1992 and 1995. What can the models be used for, and how can we know if they are any good? Let's look first at how the model can be tested.

When we constructed Model 1 we used the data from 1992 and 1993. Because we used the data from those two years to make the model, the model will work perfectly for those two years. That is, if we put $x = 3$ into the equation for Model 1 then it will produce $y = 191$, exactly the right value for y. Like this:

$$y = -6x + 209 \qquad \text{(Equation for Model 1)}$$
$$y = -6 \cdot 3 + 209 \qquad \text{(Substitute } x = 3\text{)}$$
$$y = 191 \qquad \text{(arithmetic)}$$

Similarly, if we let $x = 4$, then the correct $y = 185$ will be found. None of this should be particularly surprising, because these values determined the equation in the first place!

Now suppose we put in $x = 5$ into the equation for Model 1. The result of this calculation will be Model 1's prediction for the number of days when the State Standard was exceeded in 1994 (remember that 1994 is year 5). So, performing the calculations, we have

$$y = -6x + 209 \qquad \text{(Equation for Model 1)}$$

$$y = -6 \cdot 5 + 209 \qquad \text{(Substitute } x = 5)$$

$$y = 179 \qquad \text{(arithmetic)}$$

Model 1 predicts 179 days in 1994 when the State Standard was exceeded. The actual number of days was 165, so Model 1's prediction was 14 days too high. We can record that as an error of $+14$. Model 1's prediction for 1995 can also be found. Namely,

$$y = -6x + 209 \qquad \text{(Equation for Model 1)}$$

$$y = -6 \cdot 6 + 209 \qquad \text{(Substitute } x = 6 \text{, because 1995 is year 6)}$$

$$y = 173 \qquad \text{(arithmetic)}$$

The error here is $+19$ because the actual number of days for 1995 was 154. Table 3 shows the results of the previous test of Model 1 along with a test of Model 2.

TABLE 3   Error in Models 1 and 2

Model	Point Tested	Prediction	Actual	Error (+ high; − low)
Model 1	1994	179	165	+14
	1995	173	154	+19
Model 2	1993	178	185	−7
	1995	152	154	−2

Notice that each model was tested using the years that were *not* used in constructing that model. If the model was constructed correctly, then it should produce no error for the points that were used the construction of the model.

EXERCISE

5.1. Complete the following table.

TABLE 4   Error in Models 1–6

Model	Point Tested	Prediction	Actual	Error (+ high; − low)
Model 1	1994	179	165	+14
	1995	173	154	+19
Model 2	1993	178	185	−7
	1995	152	154	−2
Model 3	1993		185	
	1994		165	
Model 4	1992			
	1995			
Model 5				
Model 6				

You can see from Table 4 that there is quite a variety of errors in the six models. There is also a lot of variety in the predictions that they make.

6   INTERPRETING MODELS: PREDICTION

Mathematical models are used to help us understand the relationships between things and also to help us predict. Predictions are useful for planning when they are deemed valid, but they are also necessary for testing models whose validity is not yet accepted. We can use a model to make a prediction and then see whether or not the value or event predicted actually comes about. The six models we've been working with can be used to make predictions about the number of days when the State Standard was exceeded in 1996, 1997, and 1998 (none of which is given in Table 1).

To use a model to make a prediction for 1996, we simply substitute 7 (because 1996 is year 7) for x in the model's equation and then calculate the value of y. Thus, for Model 2, $y = -13 \cdot 7 + 230 = 139$ days.

EXERCISE

6.1. Complete the following table. Which model made the best prediction can be determined when the data for 1996 is available. (As I write this the data is not yet complete for 1996, but by the time you read it the data will be available at the AQMD website.)

TABLE 5   Predictions for 1996

	Prediction for Number of Days Exceeding the State Standard in 1996
Model 1	167
Model 2	139
Model 3	
Model 4	
Model 5	
Model 6	

EXERCISES

6.2. Make a table showing the 1997 predictions for each of the 6 models.

6.3. What is the prediction of Model 6 for the year 2009? How reasonable is this prediction? What factors allow us to extend the model into 2009? What factors make the extension less likely to be reasonable?

We can also use the models to make a different kind of prediction. Suppose that we wanted to know *when* there will be only 50 days exceeding the State Standard in one year. Using Model 1 and a little algebra we can answer this question. In this case, we know the y value of 50 and want to find the year x. Thus, we will substitute 50 for y in the Model 1 equation and solve for x. Like this:

$$y = -6 \cdot x + 209 \quad \text{(from Model 1 equation)}$$
$$50 = -6 \cdot x + 209 \quad \text{(substitute 50 for } y\text{)}$$
$$50 - 209 = -6 \cdot x \quad \text{(subtract 209 from both sides)}$$
$$-159 = -6 \cdot x \quad \text{(arithmetic)}$$
$$\frac{-159}{-6} = x \quad \text{(divide both sides by } -6\text{)}$$
$$26.5 = x \quad \text{(arithmetic)}$$

Our answer is that in year 26.5 the number of days when the State Standard is exceeded will total 50. Now year 26.5 would be 2015.5. Our data is not monthly data, but annual, so we are not justified in making a prediction involving anything less than a whole year. So, we'll say that in either 2015 or 2016 the number of days when the State Standard is exceeded will be 50.

EXERCISES

6.4. According to Model 2, when will the number of days when the State Standard is exceeded be 50?

6.5. According to Model 5, when will the number of days when the State Standard is exceeded be 100?

6.6. According to Model 3, when will the number of days when the State Standard is exceeded be 0?

6.7. According to Model 1, when will the number of days when the State Standard is exceeded be 0?

As we predict far into the future we have to careful. Suppose we use Model 1 to predict the ozone air quality for the year 2030. 2030 is 41 years from 1989, so Model 1 predicts that in 2030, $y = -6 \cdot 41 + 209 = -37$. Does this mean that in 2030 the South Coast Air Basin will exceed the State Standard on *negative* 37 days? Clearly, this is nonsense! The lesson here is that mathematical models can produce numbers that are worthless. It is important that every numerical value that you get from a model make sense in the context of the real-world situation that is being modeled.

EXERCISES

6.8. Is the result you got in Exercise 6.7 reasonable? What assumptions underlying the Model and the data would affect the reasonableness of the result?

6.9. Does it make sense to use the models we've constructed with the value of $x = -3$? How about $x = -100$? Explain.

6.10. Is the result you got in Exercise 6.4 reasonable? What assumptions underlying the Model and the data would affect the reasonableness of the result?

7 COMPARING MODELS: ERROR

Another aspect of mathematical model building is the task of selecting the best model from a group of models. In this module we are only concerned with simple linear models, so our "best" model will come from the list of six equations constructed in previous sections. (In addition to the linear model there is a wide variety of nonlinear models of ground-level ozone trends that we could make. You should ask your instructor about these.)

Before beginning to select the best of our six models we need to have some notion of what we mean by "best." One reasonable definition of "best model" is that the best model produces the least amount of error when tested on known data. You calculated some errors for each model when you filled in Table 4 in Exercise 5.1. A portion of that completed table is shown on p. 66.

Which of these three models has the least error? One way of calculating the total error of the model is to simply add the individual errors. Using this method, Model 1 has a total error of $+33$, Model 2 has a total error of -9, and Model 4 has a total error of $+5$. This says that Model 4 has the least total error. But wait! The total error of $+5$ for Model 4 came from the addition of a positive and a negative error. Does it make sense to let errors with

Model	Point Tested	Prediction	Actual	Error (+ high; − low)
Model 1	1994	179	165	+14
	1995	173	154	+19
Model 2	1993	178	185	−7
	1995	152	154	−2
Model 4	1992	205	191	+14
	1995	145	154	−9

opposite signs *reduce* the total error? Statisticians think not. Look at it this way. If a model had an error of +20 at one point and an error of −20 at another, would we say that the total error for the model was $+20 + -20 = 0$? Zero error means a perfect model, but a perfect model wouldn't have errors at any points, let alone an error of +20 at one point. We must conclude that adding the individual errors is not a good way to get a total error for a model.

There are two common ways of eliminating the problem of errors with different signs misleading us to a smaller than warranted total error. One way is to disregard the signs entirely. Under this method, called the *absolute error* method, the total error for Model 1 would be $14 + 19 = 33$, for Model 2 it would be $7 + 2 = 9$, and for Model 4 it would be $14 + 9 = 23$.

Another approach is to *square* the errors before adding them together. The squaring method would yield $14^2 + 19^2 = 196 + 361 = 557$ for Model 1, $(-7)^2 + (-2)^2 = 49 + 4 = 53$ for Model 2, and $14^2 + (-9)^2 = 196 + 81 = 277$ for Model 4. The numbers that arise from squaring can often get quite large, so it is customary to take the square root after we finish adding up the squared errors. Thus, for Model 1 we'll have $\sqrt{557} \approx 23.60$, for Model 2 we have $\sqrt{53} \approx 7.28$, and for Model 4 we have $\sqrt{277} \approx 16.64$. Table 6 summarizes our calculations.

TABLE 6   Total Error of Models

	Total Absolute Error	Total Squared Error
Model 1	33	23.60
Model 2	9	7.28
Model 3		
Model 4	23	16.64
Model 5		
Model 6		

You can see from Table 6 that Model 2 is the best model under both techniques because it has the least total error, but you may wonder why we should go through the squaring and square root process when the absolute error works just fine. One reason is that the squared error can often tell us more about our model. Suppose that one model has errors of 6 and −6 and another has errors of 10 and −2. Now the total absolute error for both models is 12. However, the total squared error for the first model is $\sqrt{6^2 + (-6)^2} = \sqrt{72} \approx 8.49$, but for the second model the total squared error is $\sqrt{10^2 + (-2)^2} = \sqrt{104} \approx 10.2$. The first model has a lower total squared error! This result reflects the fact that the second model had a "large" error (10) while the first model had only "medium-sized" errors (6 and −6).

EXERCISES

7.1. Complete Table 6.

7.2. Using the completed Table 6, which model is the best model? Why?

We can make one more modification to our error calculation that will give us a little better picture of how well the model actually fits the data. A total error is a nice estimate of the model's accuracy, but if we have many test points (we only have used two in our models!) this total could be quite large. If we take the average of the individual errors, then the result will be in the same ballpark as the individual errors.

We can take the average of the absolute errors by simply dividing the total absolute error by the number of data points tested. So, for Model 1 we'd get $33 \div 2 = 16.5$, for Model 2 we'd get $9 \div 2 = 4.5$, and for Model 4 we'd get $23 \div 2 = 11.5$. Model 2 still comes out as the best, but now we have an idea of how much the model is in error at each point.

When we work with the squared errors, we find the average of the squared errors and then take the square root of that average. Thus, for Model 1 we would have

$$\sqrt{\frac{14^2 + 19^2}{2}} = \sqrt{\frac{557}{2}} = \sqrt{278.5} \approx 16.69.$$

The calculation for Model 2 would be

$$\sqrt{\frac{7^2 + 2^2}{2}} = \sqrt{\frac{53}{2}} = \sqrt{26.5} \approx 5.15.$$

Finally, for Model 4 we would get

$$\sqrt{\frac{14^2 + 9^2}{2}} = \sqrt{\frac{277}{2}} = \sqrt{138.5} \approx 11.77.$$

Again, Model 2 comes out the best and we have a estimate of the error of each model at a point. This last method uses notions that are fundamental to the statistical analysis of models and is the most commonly used way of judging average error.

EXERCISES

7.3. Complete Table 7.

TABLE 7   Total Error of Models

	Average Absolute Error	Square Root of Average Squared Error
Model 1	16.5	16.69
Model 2	4.5	5.15
Model 3		
Model 4	11.5	11.77
Model 5		
Model 6		

7.4. Using Table 7 which is the best model? Why?

7.5. Use the best model to predict the number of days when the State Standard was exceeded in 1996.

7.6. Use the best model to find the number of days when the State Standard was exceeded in 1989. How did your answer compare to the actual number of days in 1989?

7.7. Use the best model to predict the year in which the number of days when the State Standard will be 30.

8   OPTIONAL EXERCISE

In actual practice, we would try to use all available data in a single, more sophisticated model than our six models, which use only two points at a time. If we had only the four data points shown in Table 1 (1992–1995) and wanted to fit the "best" straight line to those points, a program contained in many hand-held calculators will easily produce such a model. If such a calculator is available (see "regression" in your manual), carry out the following exercises:

EXERCISES

8.1. Verify that the best linear model is given by $y = 13.10x + 232.7$ (Remember to let the year 1992 correspond to $x = 3$.)

8.2. Do any of the four given data points lie on this line? How do you know?

8.3. Verify that the sum of the squares of the four errors using the above equation is 32.7. Notice that the sum of the square of these errors is less than those for Models 1, 2 and 4 (557, 53 and 277, resp.) even though we are adding the squares of all four numbers, not just two at a time. The entries in the second "Error" column in Table 6 were found by taking the square root of the sum of the square of the errors. Take the square root of 32.7 and verify that it is smaller than any of the entries in the second column.

8.4. You can do even better by taking the square root of the *average* of the squares of the four errors, as was done in Table 7. Since we have four numbers, we would divide 32.7 by 4 and then take the square root. Taking the square root of 32.7/4 is the same as dividing the square root of 32.7 by 2. So all you have to do is take half of your answer from Exercise 8.3. Verify that this averaged error is much smaller than the entries in the second error column of Table 7.

9 EXTENSION: MODELING HEALTH ADVISORY DAYS

Let's look back at Table 1. The third column gives the number of days in each year in which an ozone health advisory was issued. Health advisories are issued when the ground-level ozone reaches dangerous concentrations. Here is a plot of the health advisory data. Year 0 is 1989.

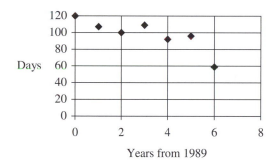

Figure 3. Health Advisory Days 1989–1995

EXERCISES

9.1. Draw a straight line connecting the dot representing the year 1989 to the dot representing the year 1994. Call this Line A.

9.2. Draw a straight line connecting the dot representing the year 1989 to the dot representing the year 1993. Call this Line B.

9.3. Which of the two lines, Line A or Line B, seems to best fit the data? Why?

9.4. Find the linear equation for Line A. Call this Model A.

9.5. Find the linear equation for Line B. Call this Model B.

9.6. Complete Table 8.

TABLE 8   Error in Models A & B

	Year Tested	Prediction	Actual	Error
Model A	1990			
	1991			
	1992			
	1993			
	1995			
Model B	1990			
	1991			
	1992			
	1993			
	1995			

9.7. Which is the better model of the data, Model A or Model B? Why?

9.8. Use the model you selected as best in the last exercise to predict the number of health advisory days in 1996. Do the same for 1997and 1998. If you have internet access you can see how well the model did by getting visiting the AQMD website.

9.9. Using the model you selected as best, predict the year in which there will be no health advisories issued. How reasonable is this prediction? What assumptions are behind this prediction?

9.10. Could the model you selected as best be used to find the number of health advisory days in 1988? How about 1970? How about 1920? Explain.

Miracles and Mathematical Biology:
The Case of the White Buffalo

Julian Fleron[1]
Donald Hoagland
Westfield State College, MA

1   INTRODUCTION

Native American legend holds that many generations ago during times of scarce game and starvation a woman appeared among the Native Americans, gave them a sacred pipe, taught them sacred ceremonies, and instructed them on the value of the buffalo. As she walked away she turned into a young white buffalo. When she returns, all humanity will be united in peace and prosperity. It is not surprising then that a female white buffalo calf born on a ranch in Wisconsin in August 1994 caused great excitement among Plains Indian tribes and others interested in Native American culture.

For thousands of years great numbers of North American bison (*Bison bison*), usually referred to as buffalo, roamed across much of the North American continent (Meagher, 1986). However, by 1903 the population had been nearly exterminated, with only 1644 remaining (Garretson, 1938). Was this drastic decline in the population size sufficient to eliminate the genes responsible for white coat color? If so, are the recent births of four

---

[1] This author would like to dedicate this paper to the fond memory of many childhood trips to the Great Plains— home of the buffalo and his maternal heritage. This paper grew out of the Brown Bag Symposium "The American Bison" sponsored by the Westfield State College Honors Program.

white buffalo calves truly miracles?[2] And, was this drastic population bottleneck sufficient to reduce or eliminate genetic variation (an important quantity for long-term survival of animal populations) in the surviving herds as has been reported in cheetah (O'Brien et al, 1983) and northern elephant seal poplations (Bonnell and Selander, 1974)?

Legends and miracles are the stuff of symbolic truth. Probabilities and frequencies are the stuff of mathematical truth. Come with us on an exploration of the probability of occurrence of a white buffalo calf, an investigation of the effects of drastic population fluctuations upon the long-term survival of buffalo, and an application of mathematics to biological problems.

2   PLACING THE BUFFALO SLAUGHTER IN PERSPECTIVE

The drastic slaughter of the buffalo during the latter part of the nineteenth century, primarily 1865–1885, caused serious biotic disturbances (Hornaday, 1889). These disturbances, in conjunction with our current reliance upon beef cattle as a food staple, continue to have serious environmental consequences (Matthews, 1990, 1992; Raven, Berg, and Johnson, 1995; Callenback, 1996).

An increased understanding of the dramatic change in magnitude of the buffalo population since the 1800s is best obtained through a comparison with familiar objects of a similar relative size. We compare the buffalo population with the current human population in the regions the buffalo inhabited.

Figure 2.1 depicts both the primary and extended range of the buffalo on the North American continent. An estimated 30 to 60 million buffalo inhabited North America at the beginning of the 19th century (McHugh, 1972; Roe, 1970). We will use the average, $N_{\text{Extended}}^{\text{Buffalo}} = 45$ million, as our estimate here. Numerical estimates for the difference in population densities between the primary and secondary ranges are not available. But let us suppose, for our current purposes, that two thirds of the population was limited to the primary range. This gives a population in the primary range of $N_{\text{Primary}}^{\text{Buffalo}} = (.66)(45,000,000) = 30,000,000$. What is the comparable human population in these ranges?

In computing the current human populations in these ranges, we include those states and provinces where the buffalo range overlaps significant human population densities and exclude those where they do not. Hence, for the primary range we have 13 states and two provinces. The total population in this range, computed from 1994 data, is $N_{\text{Primary}}^{\text{Human}} = 31,909,415$. For the extended range we have 30 states and two provinces, with a total population of $N_{\text{Extended}}^{\text{Human}} = 144,573,281$. If we compute the ratio of buffalo to human beings in these ranges we have:

$$\frac{N_{\text{Extended}}^{\text{Buffalo}}}{N_{\text{Extended}}^{\text{Human}}} = \frac{45,000,000}{144,573,281} = 0.3113$$

---

[2]In August 1994, the first white buffalo calf to survive since the birth of Big Medicine in 1933 (Geist, 1996) was born on Dave Heider's farm in rural Janesville, WI (Wronski, 1994). Three additional white calves have been born since then, but only two have survived (Foss, 1996).

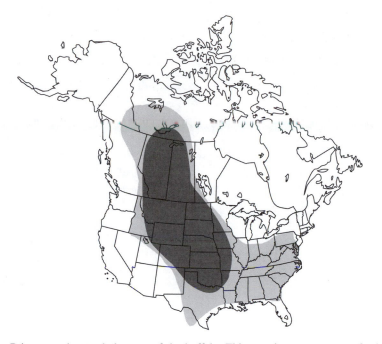

Figure 2.1. Primary and extended range of the buffalo. This map is an aggregate of other similar maps (Geist, 1996; Hodgson, 1994; McDonald, 1981; McHugh, 1972; Roe, 1970).

and

$$\frac{N_{\text{Primary}}^{\text{Buffalo}}}{N_{\text{Primary}}^{\text{Human}}} = \frac{30,000,000}{32,549,110} = 0.9217.$$

These numbers are striking. For we see that in the extended range, which includes over 80% of our country's landmass, the number of buffalo was greater than 30% of the *current* human population size. In the primary range the number of buffalo was greater than 90% of the *current* human population size! This buffalo population was eradicated, mainly, over a twenty-year period. Said differently, imagine 90% of the population in each of Arkansas, Colorado, Idaho, Iowa, Kansas, Minnesota, Missouri, Montana, Nebraska, North Dakota, Oklahoma, South Dakota, Wyoming, Alberta and Saskatchewan disappearing over a twenty-year period!

We have used human populations to illustrate the magnitude of the buffalo slaughter. We should remember that this slaughter was in fact linked to a very human tragedy—the destruction and forced relocation of numerous Native American tribes. While the genetic, biotic, and environmental consequences of the slaughter of the Native Americans cannot be as readily studied as those of the buffalo, they are of great significance, as are the cultural, historic, and moral consequences which can, and should, be readily studied.

3   PREDICTING THE FREQUENCY OF ALBINOS AND CARRIERS OF ALBINISM IN BUFFALO POPULATIONS

3.1   Introduction to Coat Color Genetics

White coat color in mammals is caused by the lack of the pigment melanin in the hair. This lack of melanin in the hair may be caused by a genetic abnormality called **albinism**. Albinism results from the inheritance of two recessive **alleles** (more commonly called genes) for the albinism trait (Klug and Cummings, 1997). The alleles are called **recessive** because two are required for the trait to be expressed. Therefore, an albino buffalo calf born to normally pigmented parents must have received one albino allele from each parent. Since the parents are not albino, each must have had one normal allele and one albino allele—such individuals are known as **carriers**. Albinos have white fur and pink eyes.

In most mammals, the pigment melanin is regarded as an evolutionary adaptation that prevents ultraviolet light damage to DNA in the cells beneath the outer layer of fur and skin. Additionally, it is generally believed that pigment in the iris of the eyes prevents damage to the retina and possibly blindness. Therefore, albinos frequently die before they reach sexual maturity. Natural selection working in this manner keeps the number of albinos in a population extremely small.

White coat color in some mammals may also be caused by a pair of recessive alleles at a different locus than albinism. Shorthorn cattle (*Bos taurus*) with brown pigmented eyes will be white if they inherit two recessive alleles for white coat color (Russell, 1990).

Because buffalo and cattle have been successfully interbred during the last 100 years (Eddy, 1996), white buffalo calves born during this century may be the result of albinism or of the expression of white coat color alleles inhereted from cattle. These two possibilities are easily differentiated, however, because only albinos have pink eyes. For the purpose of this investigation, we assume that white coat color in buffalo results from a pair of recessive alleles, causing albinism.

3.2   Quantitative Analysis of Mutation

Because natural selection eliminates detrimental alleles such as those responsible for albinism, one might ask why albinos exist at all. Similarly, one might also ask why human genetic diseases such as cystic fibrosis, sickle-cell anemia, Tay-Sachs disease, thalassemia, etc., still occur. There are actually two explanations for the continued existence of these types of genetic abnormalities. The first derives from the fact that two copies of the recessive detrimental allele must be present for the condition to be expressed; when two copies of the same allele are present, the condition is called **homozygous**. Carriers, also called **heterozygotes**, generally don't suffer; detrimental recessive alleles are maintained in the population because they are hidden from the forces of natural selection. The second explanation lies in the chemistry of DNA replication. Before a cell divides, it must replicate its entire compliment of DNA. The daughter cell receives an exact replica of the parent cell's DNA, with the exception of sperm and eggs which receive only half of the DNA. This process is one of the most efficient and exact in nature. However, mistakes are made at predictable frequencies, and are called **mutations**.

Let us consider a given genetic locus where there are two alleles that we denote by A and a. We are interested in the frequency of these alleles in a given population at a given time. Let us denote the **allelic frequency** of A by p and the **allelic frequency** of a by q. Because there are only two alleles at this locus, we will always have $p + q = 1$.

Example 3.1 Suppose we have a population of 1,000 individuals and the allelic frequency of A is $p = .87$. Because there are two alleles at each locus, there are $2 \cdot 1,000 = 2,000$ alleles in the population under consideration. Because 87% of them are A, we have $.87 \cdot 2,000 = 1,740$ A alleles in the population. As there are 2,000 total alleles, the remaining $2,000 - 1,740 = 260$ must be a alleles. And 260 a alleles in a population of 2,000 alleles gives an allelic frequency of $q = 260/2,000 = .13$. As noted, we have

$$p + q = .87 + .13 = 1.$$

We would like to study how these frequencies change under the influence of mutation alone; i.e., we assume that other evolutionary forces such as natural selection, genetic drift, and migration are not at play. Simplifying assumptions such as these are critical in quantitative models. In the current context these are reasonable assumptions. Except for the presence of mutation our population is said to be in *Hardy–Weinberg equilibrium*, a stable state described in Section 4.1.

There are two types of mutation that affect these frequencies. First, the dominant allele A can mutate into the allele a in an offspring. Such mutation will be called a **forward mutation**. On the other hand, the allele a can mutate into the allele A in an offspring. We shall call this mutation a **backward mutation**. The effect of forward mutations are to increase the frequency q of the allele a, while the effect of backward mutations are to increase the frequency p of the allele A. We would like to study these effects quantitatively.

To initiate such a study it is clearly important to know how often, i.e., at what rate forward and backward mutations occur. The more often forward mutations occur, the more q will increase. Similarly, the more often backward mutations occur, the more p will increase.

Example 3.2 Suppose we begin with a completely isolated, (AA)-homozygous population, e.g., one where $p = 1$ and $q = 0$. Because the population is (AA)-homozygous, one would expect all alleles in offspring to be A. However, because of mutation one expects to see some a alleles. Suppose there are 100 births in the next generation, and 2 a alleles are found. Because there are two alleles per birth, there are 2 mutations out of a possible 200 and it makes sense to say the forward mutation rate is $2/200 = .01$.

In this hypothetical example the founders are all (AA)-homozygous. Therefore, all a alleles that arise must arise by mutation alone. Although it becomes extremely difficult to measure mutation rates in less simplistic populations, the underlying idea is the same: of all A alleles that we expected to be passed to offspring, what percentage of them mutated to an a allele? More formally, the **forward mutation rate**[3] is

---

[3]Mathematicians and scientists often use Greek letters to denote various quantities. We will use the Greek letters mu (μ), nu (ν), beta (β), delta (δ), rho (ρ), and gamma (γ).

$$\mu = \frac{\text{number of forward mutations } A \longrightarrow a}{\text{expected number of } A \text{ alleles passed to offspring}}.$$

Similarly, the **backward mutation rate** is

$$\nu = \frac{\text{number of backward mutations } a \longrightarrow A}{\text{expected number of } a \text{ alleles passed to offspring}}.$$

While it might seem feasible that these mutation rates change from generation to generation, these mutations are ultimately caused by changes at the molecular level. Hence, it is generally believed that the forward and backward mutation rates for a given species at a given gene locus are relatively constant through time. (See Exercise 11 for further discussion.)

Example 3.3   Genetic research indicates that a typical forward mutation rate at a given gene locus is $\mu \approx 10^{-6}$ (Klug and Cummings, 1997). The United States had a population of 266,000,000 in 1996. Suppose we consider a locus in the population where an allele A has frequency that is precisely $p = .94$. We take a human generation to be 20 years. How many forward mutations might we expect in the next generation?

Define the **generational birth rate**[4] by

$$\beta = \frac{\text{number of births during the generation}}{\text{population at the beginning of the generation}}.$$

The generational birth rate for the generation 1970–1990 in the United States is $\beta = 0.4207$.[5] Assuming a similar birth rate over the generation 1996–2016, there will be approximately $(0.4207)(266,000,000) = 111,906,200$ births in this generation. This means there are 223,381,400 passed alleles. Because the allelic frequency of A is 0.94, without the effects of mutation, we expect

$$(0.94)(223,381,400) = 210,383,656 \quad A \text{ alleles passed to offspring.}$$

Hence, from the definition of the forward mutation rate we have

$$\mu = 10^{-6} = \frac{\text{number of forward mutations } A \longrightarrow a}{\text{expected number of } A \text{ alleles passed to offspring}}$$

$$= \frac{\text{number of forward mutations } A \longrightarrow a}{210,383,656}.$$

---

[4]Generally birth rates are measured over fairly short periods of time. Our model requires these rates over the course of an entire generation. For populations with rather stable birth and death rates this issue will not be problematic. However, for populations with rapidly changing birth or death rates it raises a difficult issue we will not consider. See Exercises 12 and 13 for further information.

[5]This number is computed using information from (Kurian, 1994). See Exercises 7 and 8 for details and comments.

Solving, we find the number of forward mutations $A \longrightarrow a$ is equal to

$$(10^{-6})(210,383,656) \approx 210.$$

Notice this is an exceedingly small number when compared to the total number of A alleles we expect to be passed.

One should not be misled by the previous example. Certainly, the absolute number of mutations is very small compared to the number of passed alleles. However, like the effects of compound interest, over time mutation can play an important role—as we will show in Section 3.3.

Let p_0 be the frequency of allele A at the beginning of some initial generation, and p_1 be the frequency at the beginning of the next generation. We would like to determine how these frequencies change as a result of mutation. For buffalo, humans, and other mammals the forward mutation rate, $\mu \approx 10^{-6}$, is three orders of magnitude larger than the backward mutation rate, $v \approx 10^{-9}$ (Klug and Cummings, 1997). Hence, the forward mutation rate will dominate all computations. (See Exercises 9 and 10 for an analysis of the forward and backward mutation rates simultaneously, as well as some consequences of such an analysis.) So we will consider only the forward mutation rate.

Example 3.4   Returning to Example 3.3, we had $p_0 = 0.94$. To compute p_1 we need to know the total number of alleles at the beginning of the next generation, which is simply twice the population, as well as the total number of A alleles at that point. We know there are 111,906,200 births. What about the number of deaths?

Define the **generational death rate** by

$$\delta = \frac{\text{number of deaths during the generation}}{\text{population at the beginnning of the generation}}.$$

As in Example 3.3, we can compute the generational death rate over the generation 1970–1990 in United States, yielding $\delta = 0.1970$. Hence, we can expect

$$(0.1970)(266,000,000) = 52,402,000$$

deaths. So we expect the total population at the beginning of the next generation to be: $266,000,000 + 111,906,200 - 52,402,000 = 325,504,200$. This gives a total of 651,008,400 alleles in the population. The number of A alleles at this point is determined by

1. the number of A alleles remaining from the previous generation, together with
2. the expected number of A alleles passed to offspring, with
3. the number of forward mutations $A \longrightarrow a$ removed.

We consider each of these in turn. Of the 266,000,000 alive in the first generation, we expect $266,000,000 - 52,402,000 = 213,598,000$ to survive into the next generation. As

$p_0 = .94$, we expect $(.94)(2(213{,}598{,}000)) = 401{,}564{,}200$ A alleles to survive. As we noted in the example above, we expected 210,383,656 A alleles to be passed and, of these, 210 to undergo forward mutation. Hence, at the beginning of the next generation we expect

$$401{,}564{,}200 + 210{,}383{,}656 - 210 = 611{,}947{,}646$$

A alleles. This gives an allelic frequency of

$$p_1 = \frac{611{,}947{,}646}{651{,}008{,}400} \approx .939999616.$$

Following this example, we can determine a formula that relates p_0 and p_1. Let us denote the population at the beginning of the first generation by N_0. Then the population at the beginning of the second generation is given by

$$N_1 = (N_0) + (\text{births}) - (\text{deaths})$$
$$= N_0 + \beta N_0 - \delta N_0 = N_0(1 + \beta - \delta).$$

The number of alleles at the beginning of the second generation is therefore

$$2N_0(1 + \beta - \delta).$$

The number of individuals that remain from the first generation to the next is simply $N_0 - (\text{deaths}) = N_0 - \delta N_0$. So the number of A alleles that remain from the first generation to the next is

$$2p_0(N_0 - \delta N_0).$$

The number of offspring born during the first generation is βN_0. Thus, the expected number of A alleles passed to offspring is

$$2p_0\beta N_0.$$

As $2p_0\beta N_0$ is the expected number of A alleles passed to offspring, the number of forward mutations $A \longrightarrow a$ is

$$2p_0\beta N_0\mu.$$

Putting this together, we have

$$p_1 = \frac{\text{number of } A \text{ alleles at the beginning of the second generation}}{2N_1}$$
$$= \frac{2p_0(N_0 - \delta N_0) + 2p_0\beta N_0 - 2p_0\beta N_0\mu}{2N_0(1 + \beta - \delta)}$$
$$= p_0\left(\frac{(1 - \delta) + \beta - \beta\mu}{1 + \beta - \delta}\right)$$

$$= p_0 \left(\frac{1 - \delta + \beta}{1 + \beta - \delta} - \frac{\beta\mu}{1 + \beta - \delta} \right)$$

$$= p_0 \left(1 - \frac{\beta}{1 + \beta - \delta}\mu \right).$$

The quantity $\beta/(1 + \beta - \delta)$ will appear often. We shall denote it by γ. The formula above then becomes

$$p_1 = p_0(1 - \gamma\mu).$$

The reasoning used above can be extended to give the frequency of allele A at any generation in terms of the frequency for the previous generation; namely:

$$p_t = p_{t-1}(1 - \gamma\mu).$$

Such an expression, where the value of a quantity at some stage is given as a function of the quantity during earlier stages, is called a **recurrence relation**. Quite often recurrence relations can be solved; this is one such case. For we can express p_2 in terms of p_1: $p_2 = p_1(1 - \gamma\mu)$. In turn, p_1 can be expressed in terms of p_0: $p_1 = p_0(1 - \gamma\mu)$. Utilizing this in our expression for p_2 we obtain

$$p_2 = p_1(1 - \gamma\mu)$$
$$= [p_0(1 - \gamma\mu)t](1 - \gamma\mu) = p_0(1 - \gamma\mu)^2.$$

To find p_t, carry out this same process t times, namely:

$$p_t = p_{t-1}(1 - \gamma\mu)$$
$$= [p_{t-2}(1 - \gamma\mu)](1 - \gamma\mu) = p_{t-2}(1 - \gamma\mu)^2$$
$$= [p_{t-3}(1 - \gamma\mu)](1 - \gamma\mu)^2 = p_{t-3}(1 - \gamma\mu)^3$$
$$\vdots$$
$$= [p_{t-t}(1 - \gamma\mu)](1 - \gamma\mu)^{t-1} = p_0(1 - \gamma\mu)^t.$$

Hence, for any generation t we have:

$$p_t = p_0(1 - \gamma\mu)^t.$$

3.3   Mutation and Albinism in Buffalo

Let us now use the results of the previous section to analyze the case of the white buffalo. To do so we need to determine appropriate values of the rates μ, β, δ, and γ, as well as the length of a generation. As noted above, $\mu \approx 10^{-6}$. Beyond this our approximations will be somewhat speculative for lack of definitive data. For females "sexual maturity most commonly occurs at 2 to 4 years of age." "By age 3 most males are sexually mature," but "bulls usually do not breed cows until age 6" (Meagher, 1986, p. 5). Based on such reports, we consider 5 years to be an appropriate generational length. What about the birth and death rates?

Above we showed that $N_1 = N_0(1 + \beta - \delta)$. The reasoning used to justify this expression can be applied to any generation, yielding the recurrence relation:

$$N_t = N_{t-1}(1 + \beta - \delta).$$

This recurrence relation can be solved just as the one involving p_t, the result being:

$$N_t = N_0(1 + \beta - \delta)^t.$$

This is an important result as it gives population as a function of the generation. If we define the **generational growth rate** by

$$\rho = \frac{\text{increase in population over the generation}}{\text{population at the beginnning of the generation}},$$

then, since we are assuming there is no migration, we have $\rho = \beta - \delta$. The growth equation above then becomes

$$N_t = N_0(1 + \rho)^t,$$

which has the same form as many important growth equations such as compound interest. Equations of this type assume a constant growth rate. While that is not the situation here, especially considering the efforts of groups like the Nature Conservancy to reintroduce the buffalo to the Great Plains, it is a fair approximation given the available data and our purposes.

We can approximate the generational growth rate ρ, thus $\beta - \delta$, using known data. We let 1903 be generation 0, so we have $N_0 = 1644$. 1998 is

$$\frac{1998 - 1903}{5} = \frac{95}{5} = 19$$

generations later, and an estimate of the current population is $N_{19} = 200{,}000$ (Hodgson, 1994). Using these values the equation above becomes $200{,}000 = 1644(1 + \rho)^{19}$. Solving this for ρ gives

$$\rho = \left(\frac{200{,}000}{1644}\right)^{1/19} - 1 = 0.2875.$$

To compute γ we need β and δ individually. Recent data from the Yellowstone herd (John Mack, personal communication) allowed us to calculate eight generational birth rates whose average was $\beta = 0.5098$. Although we do not suppose this measurement generalizes to the entire population throughout this century, it does give us at least an approximate idea of a realistic value. Moreover, since small changes in the value of β do not have a dramatic effect on the value of γ (see Exercise 17), this value will suffice. Then, since $\delta = \beta - \rho$, we have $\delta = 0.5098 - 0.2875 = 0.2223$. This gives

$$\gamma = \frac{0.5098}{1 + 0.5098 - 0.2223} = 0.3960.$$

Let us now use this information to quantitatively analyze the appearance of the white buffalo.

Case 1.

It is possible that there were no a alleles for albinism left after the slaughter of the buffalo. If the remaining population in 1903 is considered the first generation 0 we have $p_0 = 1$. How has mutation changed this frequency in the intervening 95 years? As there have been $95/5 \approx 19$ generations, one would expect the frequency of allele A in 1998 to be:

$$p_{19} = p_0(1 - \gamma\mu)^{19}$$
$$= 1(1 - 0.3960 \times 10^{-6})^{19}$$
$$\approx 0.99999248.$$

Hence, $q_{19} = 1 - p_{19} \approx 1 - 0.99999248 = 7.52 \times 10^{-6}$. As we noted before, to be born with albinism, a buffalo must receive two a alleles. The probability of receiving one is simply q, so the probability of receiving two is q^2. This probability in the upcoming generation is

$$(q_{19})^2 \approx (7.52 \times 10^{-6})^2 = 5.66 \times 10^{-11}.$$

This is an exceedingly small probability. However, this is the probability that a single buffalo calf will be born white. We have $N_{19} = 200{,}000$ so we expect $\beta N_{19} = 0.5098(200{,}000) = 101{,}960$ births in the next generation. Hence, the expected number of white buffalo born in this generation is

$$(101{,}960)(5.66 \times 10^{-11}) = (1.0196 \times 10^5)(5.66 \times 10^{-11}) \approx 5.77 \times 10^{-6}.$$

Stated differently, the probability of a white buffalo being born during the upcoming generation is about 5 out of one million. This would suggest that the recent births truly are miraculous.

Case 2.

It is possible, perhaps even probable, that the genetic pool of the buffalo that survived the slaughter was not (AA)-homozygous. The buffalo had been roaming North America for thousands of years, certainly ample time for mutation to develop a significant proportion of a alleles for albinism. Additionally, because most mammals exhibit albinism, it's likely that the ancestors of *Bison bison* had albino alleles that *Bison bison* inherited. We explore the former possibility here, leaving the important later possibility for Exercise 18.

Bison bison were known to inhabit North America some 6,000 years ago (McDonald, 1981). This constitutes $6,000/5 = 1,200$ generations. What could we expect the allelic frequencies to be then during contemporary times? Beginning with an initial (AA)-homozygous population, $p_0 = 1$, the frequency of allele A 1,200 generations later is expected to be

$$p_{1200} = p_0(1 - \gamma\mu)^{1200}$$
$$= 1(1 - 0.3960 \times 10^{-6})^{1200}.$$

Evaluating such expressions can be troublesome because of significant digit limitations and roundoff error inherent in many calculators and computers. Our direct electronic calculation yielded

$$p_{1200} = 1(1 - 0.3960 \times 10^{-6})^{1200}$$
$$\approx 0.99952,$$

a result confirmed by a binomial approximation. (See Exercise 16 for discussion.) Hence, $q_{1200} = 1 - p_{1200} \approx 1 - 0.99952 = 4.8 \times 10^{-4}$. If this frequency survived through the genetic bottleneck caused by the slaughter, then, as we computed above, the expected number of white buffalo in the next generation would be

$$(101,960)(q_{1200})^2 \approx (1.0196 \times 10^5)(4.8 \times 10^{-4})^2 = 0.023.$$

This is dramatically different than the result above. For in this setting there is about a one out of forty probability that a white buffalo will be born in the next generation. Although surprising, no longer does the birth of such a legend seem so miraculous.

Case 3.

Since the time of the slaughter, there have been at least five white buffalo born. As noted above, it's quite possible that they occurred due to cross-breeding with cattle at some point in their genealogy. However, at least two have been documented albinos, and possibly four of the five are albino. This seems to be at odds with cases 1 and 2 that suggest the expected number of white buffalo born in contemporary generations is fairly small. We'd like to explore this situation by equating an expression representing the expected number of white buffalo born over the past 19 generations with what we'll assume are four albino buffalo born during this time. Before proceeding, we should note that more sophisticated statistical analyses are generally employed in questions such as this, but the reliability of these methods would be similar to that of the method below because the number of albino births in this century is so small.

As utilized in case 2, the expected number of white buffalo born in the tth generation is given by $\beta N_t(q_t)^2$. We know N_t can be expressed as $N_t = N_0(1 + \rho)^t$. So the expected number of white buffalo born in the tth generation is given by

$$\beta N_t(q_t)^2 = \beta N_t(q_0(1 - \gamma\mu)^t)^2$$
$$= \beta(N_0(1 + \rho)^t)(q_0(1 - \gamma\mu)^t)^2.$$

Hence, the expected number of white buffalo since the end of the slaughter, i.e., from 1903 through 1998, is given by

$$W = \beta(N_0(1 + \rho)^1)(q_0(1 - \gamma\mu)^1)^2 + \beta(N_0(1 + \rho)^2)(q_0(1 - \gamma\mu)^2)^2$$
$$+ \cdots + \beta(N_0(1 + \rho)^{19})(q_0(1 - \gamma\mu)^{19})^2$$
$$= \beta N_0 q_0^2 (1 + \rho)(1 - \gamma\mu)^2$$
$$\cdot \left[\begin{matrix} 1 + ((1 + \rho)(1 - \gamma\mu)^2) + ((1 + \rho)(1 - \gamma\mu)^2)^2 \\ + \cdots + ((1 + \rho)(1 - \gamma\mu)^2)^{18} \end{matrix} \right].$$

If we denote $(1+\rho)(1-\gamma\mu)^2$ by x, then the term in brackets has the form $1+x+\cdots+x^{18}$, which is the partial sum of a *geometric series*. Applying the equation from Exercise 9(d), the result is:

$$W = \beta N_0 q_0^2 (1 + \rho)(1 - \gamma\mu)^2$$
$$\cdot \left[\begin{matrix} 1 + ((1 + \rho)(1 - \gamma\mu)^2) + ((1 + \rho)(1 - \gamma\mu)^2)^2 \\ + \cdots + ((1 + \rho)(1 - \gamma\mu)^2)^{18} \end{matrix} \right]$$
$$= \beta N_0 q_0^2 (1 + \rho)(1 - \gamma\mu)^2 \left[\frac{1 - \left[(1 + \rho)(1 - \gamma\mu)^2\right]^{19}}{1 - (1 + \rho)(1 - \gamma\mu)^2} \right].$$

We've estimated all of the rates involved, and have seen $W = 4$ albino buffalo during this time. So we solve for q_0 obtaining

$$q_0 = \sqrt{\frac{W(1 - (1 + \rho)(1 - \gamma\mu)^2)}{N_0\beta(1 + \rho)(1 - \gamma\mu)^2 \left(1 - \left[(1 + \rho)(1 - \gamma\mu)^2\right]^{19}\right)}}$$

$$= \sqrt{\frac{4(1 - (1 + 0.2875)(1 - 0.3906 \times 10^{-6})^2)}{1644(0.5098)(1 + 0.2875)(1 - 0.3906 \times 10^{-6})^2}}$$

$$\cdot \sqrt{\frac{1}{\left(1 - \left[(1 + 0.2875)(1 - 0.3906 \times 10^{-6})^2\right]^{19}\right)}}$$

$$\approx \sqrt{8.8316 \times 10^{-6}}$$

$$= 2.9718 \times 10^{-3}.$$

This is the allelic frequency that would be necessary in 1903 to give an expected number of 4 albino buffalo over the next 19 generations. Is this feasible? Well, there were 1644 surviving buffalo in 1903, hence 3288 alleles. To have an allelic frequency of 2.9718×10^{-3} we would have to have $(2.9718 \times 10^{-3})(3288) = 9.7712$ surviving a alleles. In other words, we would have needed approximately 10 heterozygotes in the 1903 population of 1644. In Section 4 we'll see that this figure is consistant with measured levels of "heterozygosity."

4   PREDICTING THE IMPACTS OF POPULATION BOTTLENECKS UPON GENETIC VARIATION

4.1   Introduction to Genetic Variation

It is a well-known fact that all environments change through time: Earth is no longer covered with large tracts of tree fern and cycad forests that were home to our famous, but now extinct, Jurassic Park inhabitants. Populations respond to environmental changes in one of two ways: they go extinct, as did the dinosaurs, or they change to meet new environmental demands. In order for a population, or species, to adapt to a new environment it must possess genetic variation. In other words, populations exhibiting some genetic variation are more likely to survive environmental fluctuations than are populations containing no genetic variation.

Biologists have developed numerous techniques for estimating genetic variation in extant populations of animals and plants. A common measure is **heterozygosity**, denoted by H, which is defined as the frequency of heterozygous individuals at a given locus. Heterozygosity can be estimated from allele frequencies (e.g., p and q) through the use of the binomial expansion $(p + q)^2 = p^2 + 2pq + q^2$ under suitable conditions described below. For $p + q = 1$, we have:

$$1 = (p + q)^2$$
$$= p^2 + 2pq + q^2.$$

In Section 3 we noted that q^2 represented the frequency of (aa)-homozygotes. Interpreting each of the quantities in this way we have

$$p^2 = \text{the frequency of } (AA)\text{-homozygotes}$$
$$q^2 = \text{the frequency of } (aa)\text{-homozygotes}$$
$$2pq = \text{the frequency of heterozygotes.}$$

Consequently we have

$$H = 1 - (p^2 + q^2).$$

Biologists term the above model **Hardy–Weinberg equilibrium**, a model first formulated in 1908 independently by the mathematician G.H. Hardy in England and the physician Wilhelm Weinberg in Germany. This model can be used to predict genotype frequencies—the actual extant frequencies of a given genotype (i.e. Aa) in a population—from allele frequencies—the actual extant frequencies of alleles (i.e. A) in a population—providing the following conditions are met: the population size must be large, mating must be random, there must be no mutation, no migration, and no natural selection. Within this model, the allele frequencies and genotype frequencies will be precisely linked and will not change from generation to generation. If the population size is large (say $N > 50$), then these conditions are likely to be met for some individual genetic loci. However, not all populations are large, and one or more of the forces identified above may impact a single genetic locus if the period of time being evaluated spans many generations. For example, as discussed in Example 3.3, mutation rates are very low and therefore are generally believed to have little impact in structuring gene pools. Genetic drift is also considered to be a weak force in relatively large populations; however, in small populations (e.g., $N < 10$) these effects can be dramatic. Natural selection and nonrandom mating can cause substantial changes in a gene pool. However, recessive traits such as albinism will never be eliminated from a population because the majority of the alleles are hidden from selection (i.e., they are present in the heterozygous condition.)

A reasonable estimate of the total amount of genetic variation present in a population or species can be obtained by averaging the per locus heterozygosity over all genetic loci sampled. Proteins are representative measurements of genetic variation because they are the products of alleles. Average heterozygosity over all loci for 200 species of mammals is approximately 0.04 (Nevo, Beiles, and Ben-Shlomo, 1984).

Example 4.1   10 heterozygotes in a population of 1644, the number expected from Case 3 of Section 3, gives a heterozygosity of

$$H = \frac{10}{1644} \approx 0.0061.$$

As noted, the average of 0.04 for mammals is over all loci. We expect some fluctuations— 0.0061 is within an appropriate range.

Example 4.2   If we use the average 0.04 as an actual measure of the heterozygosity in the population of 1644 buffalo that survived the bottleneck, then we would expect

$$N \cdot H = (1644)(0.04) = 65.76$$

heterozygotes to have survived the population bottleneck. This number is significant because, if appropriate, it suggests that we should have seen even more white buffalo than we have. (See Exercise 29.)

4.2   Influence of Population Bottlenecks Upon Genetic Variation as Estimated by Heterozygosity

We would like to consider the impact of drastically reducing a population from one generation to the next, an event that is known as a **population bottleneck**, upon levels of heterozygosity. We begin with an example.

Example 4.3   Consider an asexual population that dies after it reproduces and has a population that is 25% (*AA*)-homozygous, 50% heterozygous, and 25% (*aa*)-homozygous.

Clearly we have $p = q = 0.5$, and then one easily shows that $H = 0.5$ in this situation. What is the heterozygosity after one generation? Excluding mutation, all offspring with a homozygous parent are homozygous of the same type. The 50% of offspring that are born to a heterozygous parent have a $p^2 = 0.25$ probability of being born (AA)-homozygous and a $q^2 = 0.25$ probability of begin born (aa)-homozygous. Hence, the second generation will be $25\% + (0.5) \cdot (25\%) = 37.5\%$ (AA)-homozygous, 37.5% (aa)-homozygous, and only $100\% - 2(37.5\%) = 25\%$ heterozygous. Hence, the original heterozygosity, H_0, is reduced by half in the course of one generation. In fact, it is not hard to show that after t generations the heterozygosity of the population described in this example will be given by

$$H_t = (0.5)\left(\frac{1}{2}\right)^t.$$

The case in which a species reproduces asexually provides an extreme example of a population bottleneck in terms of the effect on heterozygosity. For in such a species each individual makes up its own isolated gene pool, and, without outside effects, heterozygosity will decrease rapidly.

In the case of sexual reproduction the analysis is more complicated. In the setting of a constant-sized breeding population (see Exercise 25 for the nonconstant case) this fairly complicated analysis leads to a simple recurrence relation, $H_t = H_{t-1}(1 - (\frac{1}{2}N))$ which, when solved as we've solved the other recurrence relations, yields

$$H_t = H_0 \left(1 - \frac{1}{2N}\right)^t,$$

where H_t is the heterozygosity at generation t and N is the constant size of the breeding population (Hartl, 1980). This model assumes approximately equal numbers of males and females, a common occurrence in the animal world. Notice that in the case of asexual reproduction the breeding population size is essentially $N = 1$ since each individual breeds in isolation. Hence, the equation in Example 4.3 is a special case of this equation, which describes a more general process.

Example 4.4 In the case of the buffalo the minimum post-bottleneck population was 1644. The percentage of heterozygosity retained in the next generation was:

$$\frac{H_1}{H_0} = \left(1 - \frac{1}{2(1644)}\right) \approx 99.97\%.$$

Example 4.5 The buffalo rebounded in size fairly quickly. However, suppose they remained at a constant population of $N = 1644$ for several generations. The percentage of heterozygosity retained after 5 and 50 generations would be, respectively:

$$\frac{H_5}{H_0} = \left(1 - \frac{1}{2(1644)}\right)^5 \approx 99.85\%$$

$$\frac{H_{50}}{H_0} = \left(1 - \frac{1}{2(1644)}\right)^{50} \approx 98.49\%$$

It is apparent from the examples above, with a minimum population size of 1644, buffalo did not likely suffer a loss of genetic variation. And in fact, data published by Berger and Cunningham (1994) indicate that buffalo exhibit average levels of heterozygosity (approximately .03) similar to the average for mammals (.04).

5   CONCLUSION

The birth of a white buffalo calf can certainly be considered miraculous for spiritual reasons. However, a mathematical analysis reveals that this phenomenon is also a predictable event. The reason events such as these take on spiritual meaning is due to their very rare nature. Other rare events such as solar and lunar eclipses have been used by mystics for spiritual reasons for hundreds of years—these events are now well understood and predicted by scientists. Conceptually, these events are also understood by interested nonspecialists.

Conservation biologists use mathematical analyses similar to those presented in this paper to evaluate the status of rare and endangered species. Fortunately, the prognosis for the American buffalo is good. The population size after the slaughter was sufficient to retain sufficient genetic variation that will likely ensure the survival of the buffalo for many generations. This prospect did not, however, come about by pure luck. It was intended. The buffalo that remained after the massacre were scattered in small herds of 5–20 individuals throughout North America (Hodgson, 1994). Breeders recognized the danger of inbreeding and the probability of losses of genetic variation with such small numbers. Consequently, buffalo were transported from herd to herd for breeding purposes. Thanks to those insightful conservationists, and the continued use of sound conservation practices, we might once again witness massive herds of buffalo rumbling across the Great Plains, and maybe even have peace on Earth as prophesied by the white buffalo legends.

EXERCISES

1. A widely-used photograph (see e.g., (Geist, 1996) and (Hodgson, 1994)) gives us another way to begin to understand the magnitude of the buffalo slaughter. The photograph pictures two men standing by a giant pile of buffalo skulls that had been collected to be used for industrial purposes from the "vast boneyard" that the Great Plains became during the slaughter. The conical pile of bones is approximately 20 feet high and has a radius of approximately 30 feet.

    (a) Compute the volume of such a conical pile.

    (b) Using the fact that buffalo skulls have dimensions of approximately 1 foot by 2 feet by 1 foot, estimate the total number of buffalo skulls in the conical pile described above. (See Chapter 22 of (McHugh, 1972): "On the way to extinction" for a disturbing account of the number of buffalo hides and the volume of buffalo bones that were shipped in the 1870's.)

2. The slaughter of the Native Americans was briefly mentioned in the text.

   (a) In 1876, as Congress attempted to pass a bill to stop the indiscriminate slaughter of the buffalo, Representative James Throckmorton of Texas remarked, "There is no question that, so long as there are millions of buffaloes in the West, so long the Indians cannot be controlled, even by the strong arm of the Government. I believe it would be a great step forward in the civilization of the Indians and the preservation of peace on the border if there was not a buffalo in existence" (Hodgson, 1994). Discuss the linked plights of the Native Americans and the buffalo during the latter part of the nineteenth century.

   (b) In the text we used current human populations in the regions formerly inhabited by the buffalo to describe the magnitude of their slaughter. Find comparable data that give the number of Native Americans that were killed during the latter part of the nineteenth century. Can you make similar comparisons that describe the magnitude of their slaughter?

3. The title of the second section of this paper is "Placing the Buffalo Slaughter in Perspective."

   (a) Many people have a hard time grasping the magnitude of large numbers, including perhaps the slaughter of 45 million buffalo. Name several groups that consists of (i) thousands of objects, (ii) millions of objects, (iii) billions of objects, and (iv) trillions of objects.

   (b) Discuss the relative size differences in the objects considered in (a).

   For discussion of these kinds of issues see the captivating, controversial, bestseller *Innumeracy: Mathematical Illiteracy and its Consequences* (Paulos, 1988).

4. Suppose a population with $\mu \approx 10^{-6}$ is purely (AA)-homozygous at a specific point in time. Suppose the generational birth rates and death rates are 0.3 and 0.2 respectively. How many generations must pass before

   (a) $p_t = 0.9$?

   (b) $p_t = 0.75$?

   (c) $q_t = 0.4$?

   (d) $p_t = q_t$?

5. (a) Explain why, in the absence of backward mutation, a population will evolve toward one that is (aa)-homozygous, even if it begins as (AA)-homozygous.

   (b) Find out whether this is what is observed in long-surviving species.

   (c) What do (a) and (b) tell you about the relative importance of genetic drift, migration, mutation, and selection in determining genetic traits?

6. (a) Explain why the relationship $p + q = 1$ must always hold for the allelic frequencies p and q at a locus with two alleles.

(b) Suppose a genetic locus had three alleles and their frequencies were denoted by p, q, and r, respectively. Is there a relationship between p, q, and r that is analogous to the relationship $p + q = 1$ that we considered in (a)? Explain.

(c) Can you extend the relationship in (b) to a locus where there are n alleles whose allelic frequencies are denoted by $p_1, p_2, \ldots, p_n$, respectively? Explain.

(d) Consider how difficult it might be to derive a formula expressing a given frequency p_j over time. What if you were tracking several different loci that each had several different alleles? What does this tell you about the complexity of population genetics?

7. The population in 1970 was 203,235,298 and in 1990 was 248,709,873. Compute the generational birth and death rates for the generation 1970–1990 for the United States as follows. (The data above and below is from (Kurian, 1994).)

(a) The number of reported deaths during this generation was 40,036,000. Calculate the generational birth rate for this generation.

(b) The number of reported births during this generation was 71,264,000. Is this number of births sufficient to account for the population growth reported in this period? Explain.

(c) There were a reported 11,832,000 immigrations into the United States during the generation under consideration. Additionally, methods of calculating census data changed several times. To account for these and other discrepancies, let us agree to adjust the number of births so that they account for the entire population growth needed to account for the data above. (See Exercise 8 for further discussion.) Under this assumption, what will our adjusted number of births be?

(d) Calculate the generational birth rate from using the value from (c).

8. Discuss the implications of our assumptions in Exercise 7. In particular:

(a) Is it appropriate to include immigrants as new births when studying allelic frequencies or some other type of inherited genetic trait in a population?

(b) The U.S. Census Bureau approximates 133,000 emigrations per year, a figure that should possibly be as high as 195,000 per year (Ahmed and Robinson, 1994). Over a generation, how large will the effect of emigration be in the calculation of generational birth rates and death rates?

(c) Both immigration and emigration are examples of genetic migration. Explain, using this and the previous problem, how migration complicates quantitative questions in population genetics.

9. Derive a formula for p_t when both the forward and backward mutations are considered as follows:

(a) Following Example 3.4 we derived a recurrence relation for p_t. Adapt this derivation to include backward mutations as well, yielding a recurrence relation for p_t in terms of p_{t-1} and q_{t-1}.

(b) Use the fact that $p_t + q_t = 1$ to show that we can write

$$p_t = p_{t-1}(1 - \gamma(\mu + \nu)) + \gamma\nu.$$

(c) Use the recurrence relation in (b) to show that

$$p_t = p_0(1 - \gamma(\mu + \nu))^t$$
$$+ \gamma\nu \left[(1 - \gamma(\mu + \nu))^{t-1} + (1 - \gamma(\mu + \nu))^{t-2} \right.$$
$$\left. + \cdots + (1 - \gamma(\mu + \nu)) + 1 \right].$$

(d) A series $1 + x + x^2 + \cdots + x^n + \cdots$ is called a **geometric series**. One important result about "partial sums" of such series, which can be readily checked by "cross-multiplying," is that $1 + x + x^2 + \cdots + x^n = (1 - x^{n+1}/1 - x)$. Use this fact, and the equation from (c) to show that

$$p_t = p_0(1 - \gamma(\mu + \nu))^t + \frac{\nu}{(\mu + \nu)} \left[1 - (1 - \gamma(\mu + \nu))^t \right].$$

10. There are several important consequences of the result in 9(d).

(a) Compare the values of p_t obtained in the text with their analogues that one obtains using the formula from 9(d). Show how this comparison justifies our assumption that the backward mutation rate can be ignored.

(b) Show that, in the absence of any other evolutionary forces, the frequency p_t tends towards $\nu/\mu + \nu$ as a given lineage continues to evolve over time.

(c) What is this limiting value, $\nu/\mu + \nu$, for mammals? What does this tell you about the evolution of mammals in the absence of evolutionary factors other than mutation?

(d) What do (b) and (c) tell you, in general, about the relative importance of genetic drift, migration, mutation, and selection in determining the genetic structure of populations through time?

11. In the text we noted that the forward and backward mutation rates are generally assumed to be constant over time. However, environmental factors certainly can play a role in these rates. Name some environmental factors that you think might change these rates of mutations.

12. In this problem you are asked to adapt Examples 3.3 and 3.4 to populations where the birth and death rates are not stable. Suppose an isolated population has a generational length of 10 years. Further suppose that population, the number of births and the number of deaths over the past 10 years are as given, in thousands, in the following tables.

	1986	1987	1988	1989	1990	1991	1992	1993	1994	1995
Population	203	203	200	197	192	196	210	225	237	253
Births	14	12	14	11	18	24	26	21	23	18
Deaths	14	15	17	16	14	10	9	9	7	8

(a) Determine the generational birth rate for the ten-year period 1986–1996.

(b) Determine the generational death rate for the ten-year period 1986–1996.

(c) If we used these rates to predict the number of births, number of deaths, and population growth during the next generation, 1996–2006, how reliable would the results be? Explain.

(d) What kinds of events might cause a population to have fluctuating birth and death rates such as those indicated in the table above?

13. An immediate reaction to address the difficulty described in Exercise 12 is to make all calculations over shorter time periods, say one-year periods rather than generations. All of the constants and formulas derived above can then be suitably adjusted. In particular,

$$p_t = p_0(1 - \gamma\mu)^t,$$

where t is now measured in years. However, such models have one-year old children having children. Explain.

14. Most texts that discuss population genetics assume that populations are nonoverlapping. That is, one generation is born, bears offspring, and then dies before the next generation reproduces. This assumption simplifies the nature of the mathematical models that are used.

(a) Determine an appropriate value for δ when there are nonoverlapping generations.

(b) Use (a) to describe how our more sophisticated model for predicting allelic frequency, $p_t = p_0(1 - \gamma\mu)^t$, can now be adapted to the simpler model $p_t = p_0(1 - \mu)^t$, that appears in most texts.

(c) Describe some populations where the assumptions above would be appropriate and would make the formula in (b) robust.

(d) Describe some populations where the assumptions above would not be appropriate and would adversely effect the robustness of the formula in (b).

15. Redo the final calculations in Cases 2 and 3 of Section 3 under the simplifying assumption of nonoverlapping generations, as in Exercise 14. Compare your answers to those we obtained in the text.

16. In many of these cases we are using very large numbers, very small numbers, and often very high powers. It is important to be wary of significant digits and calculator round-off. What follows describes a way to evaluate $(1 - \gamma\mu)^t$ without having to compute it directly.

(a) Expand $(1 - x)^2$.

(b) Expand $1 - x)^3$.

(c) You might recognize the pattern of the coefficients. They come from what is called "Pascal's triangle," which gives the "binomial coefficients" for such expansions.

In general one has:

$$(1 - x)^n = 1 - nx + \binom{n}{2}x^2 - \binom{n}{3}x^3 + \cdots + (-1)^n \binom{n}{n}x^n.$$

Letting $x = \gamma\mu$, use this formula to express $(1 - \gamma\mu)^t$.

(d) For $\gamma\mu$ extremely small, as is often the case, explain how important the terms involving $(\gamma\mu)^2, \ldots, (\gamma\mu)^t$ will be in the formula from (c).

(e) Conclude that $(1 - \gamma\mu)^t \approx 1 - t\gamma\mu$.

(f) Compare the result you obtain using $1 - t\gamma\mu$ with that in the text.

17. One can use the approximation from part (e) of Exercise 16 to understand how much changes in the parameters β, δ, ρ, and γ have on our calculation of allelic frequencies. Such a study is usually called "sensitivity analysis."

(a) For our analyses of buffalo we were able to determine the generational growth rate $\rho = \beta - \delta$ fairly precisely over the period 1903–1998. However, we were unable to determine β or δ accurately. How much will our computation of allelic frequencies change if an appropriate value of δ is .30 rather than .2223 as we have been using? If it is .20?

(b) The value of γ we used above is $\gamma = 0.3960$. The simplified setting of Exercise 14 essentially results in a value of $\gamma = 1$. Use the approximation above to address how significant of a difference this results in when computing allelic frequencies.

18. *Bison bison* evolved from earlier bison, including *Bison antiquus*, *Bison latifrons*, and *Bison priscus*. It is the latter that is thought to have crossed from Siberia to Alaska during the Ice Age. (See (Geist, 1996) for discussion.) *Bison latifrons* is known to have inhabited North America 300,000 years ago. As *Bison bison* descended from this line, it can be argued that our calculation of allelic frequency for albinism should have started at this point.

(a) Compute the allelic frequency q in contemporary times assuming a (AA)-homozygous population 300,000 years ago.

(b) Compute the probability of a white buffalo calf being born in the generation starting in 1998 assuming the frequency you found in (a) survived the genetic bottleneck of the 1880's.

(c) No white buffalo are known to have survived the genetic bottleneck. How many heterozygous buffalo would have to have survived to carry on the frequency you determined in (a)?

(d) Compare your answers here with the results in Cases 1–3. In particular, what are your thoughts now in regard to the "miraculous" birth of white buffalo?

19. Broadly speaking, the issue of the number of heterozygotes that remained through the genetic bottleneck is what is considered genetic drift, a topic that is considered in Section 4 of this chapter. In a population as small as 1644, one or two more heterozygotes surviving than expected, given the allelic frequency before the slaughter, can have a fairly large effect on the genetic structure of subsequent populations.

(a) Suppose that one heterozygote remained in the population of 1644 in 1903. Compute the allelic frequencies in 1998 as well as the expected number of white buffalo born in the generation beginning 1998 based on these frequencies.

(b) Suppose that three heterozygotes remained in the population of 1644 in 1903. Compute the allelic frequencies in 1998 as well as the expected number of white buffalo born in the generation beginning 1998 based on these frequencies.

(c) Suppose that five heterozygotes remained in the population of 1644 in 1903. Compute the allelic frequencies in 1998 as well as the expected number of white buffalo born in the generation beginning 1998 based on these frequencies.

(d) Compare these results.

20. In the text we based our heterozygosity calculations on the entire population of 1644. In fact, the buffalo population that remained in 1903 had been segregated into smaller groups: Yellowstone National Park, the Bronx Zoo, the National Zoological Park, etc. Use the formula $H_t = H_0(1 - (\frac{1}{2}N))^t$ to answer the following questions.

(a) In a segregated population founded with 25 individuals, what percentage of the existing heterozygosity would be retained after one generation? After 5 generations?

(b) In a segregated population founded with 12 individuals, what percentage of the existing heterozygosity would be retained after one generation? After 5 generations?

(c) What do these examples suggest about the loss of heterozygosity in the buffalo due to a small number of segregated herds existing in the early 1900's?

21. (a) Find a population size in which after 2 generations less than 70% of the original heterozygosity is retained.

(b) Is there a population size in which after 2 generations less than 60% of the original heterozygosity is retained?

(c) Is there a population size in which after 2 generations less than 50% of the original heterozygosity is retained?

22. Compute the heterozygosity for each Case 1–3 in Section 3.3 using the allelic frequencies that arise at the end of the time periods under consideration. Assume that the population is in Hardy–Weinberg equilibrium.

23. (a) For each of the results from Exercise 22, compute the heterozygosity retained 2 generations after the bison population was reduced to $N = 1644$.

(b) Repeat (a) assuming a segregated population of 15 individuals.

(c) Contrast your results; what are the environmental implications?

24. In small, isolated populations there is a real possibility that inbreeding will adversely effect the level of heterozygosity. Assume that a population of size $N = 15$ has a constant size and is isolated from outside genetic influences.

(a) Determine the number of generations that must pass before less than 75% of the original heterozygosity is retained.

(b) Determine the number of generations that must pass before less than 50% of the original heterozygosity is retained.

(c) If the population retains its original size, and continues to be isolated from outside genetic influences, what does our formula suggest about the long term retained level of heterozygosity?

25. Our formula for heterozygosity assumes a constant population. Nonetheless, this formula can be adapted to the nonconstant case in the following way.

(a) Use the formula to express H_1 in terms of H_0 given an initial population of N_0.

(b) Use the formula to express H_2 in terms of H_1 given an initial population of N_1.

(c) Use (a) and (b) to find an expression for H_2 in terms of H_0, N_0, and N_1.

(d) Explain how the reasoning used in (c) can be extended to find an expression for H_t in terms of $H_0, N_0, N_1, \ldots, N_t$.

(e) Under the assumption of a constant growth rate, we have seen that we can express any N_t as a function of N_0, β, δ, and t. Explain how this can be used to simplify the expression in (d).

26. Use Exercise 25 and the generational growth rate for buffalo used in Section 3 of the text to calculate the percentage of heterozygosity retained in a genetically isolated herd after

(a) five generations given a founding population of $N = 10$.

(b) twenty generations given a founding population of $N = 10$.

(c) five generations given a founding population of $N = 6$.

(d) twenty generations given a founding population of $N = 6$.

27. Exercise 18 addresses the important case when the allelic frequency q is computed beginning with *Bison latifrons*. Using the results of this problem, complete the following.

(a) Compute the retained heterozygosity one generation after the population reached 1644.

(b) Compute the retained heterozygosity three generations after the population reached 1644.

(c) What do these results suggest about the possibility of retaining heterozygosity after a genetic bottleneck?

(d) Do the results here suggest that any substantive adjustment in your answer to Exercise 18(d) is necessary?

28. Given the allelic frequency determined in Exercise 18(a), consider the effects on heterozygosity given a small, isolated founding population with $N = 4$ as follows.

(a) Compute the retained heterozygosity one generation after the population became isolated.

(b) Compute the retained heterozygosity three generations after the population became isolated.

(c) These answers should give you more evidence that the extant heterozygosity at the time of the slaughter of the buffalo was, to a significant degree, retained. Explain.

29. Example 4.2 suggests that there may have been as many as 66 heterozygotes in the population of 1644 that survived the slaughter.

   (a) Compute the allelic frequency q_0 in 1903 given that there were no white buffalo among the 1644 that survived.

   (b) Compute the allelic frequency q_{19} that would result 19 generations later.

   (c) Use (b) to compute the expected number of white buffalo in the generation 1998–2003.

   (d) Is this surprising?

30. Native American culture is not the only culture that has legends of white coated animals.

   Southern Chinese culture has a tale of a white water buffalo. In the tale, a young orphan draws a picture of a white water buffalo—"the rarest of all creatures." In the morning the picture has come to life. Unfortunately, the white buffalo is stolen by a warlord. Despondent, the child draws a dragon in the hope that it could aid him in freeing the white water buffalo. In the morning the dragon has indeed come to life and helps the child liberate the white buffalo. The three friends then fly away to the mountains.

   One might expect revered mammals in any culture to give rise to such legends revolving around white coat coloring.

   (a) See if you can find such a legend other than the native American and Southern Chinese legends that have been considered here. Describe this legend, give some idea what role the animal played in the culture, and consider how likely the appearance of such a creature might be using some of the lessons that have been learned through our investigations above.

   (b) If you are unable to find a different legend, create your own. That is, learn about the culture of a given people, consider a mammal that is important to that people and then create a legend that would seem appropriate for such a culture. Discuss how likely the appearance of such a creature might be using some of the lessons that have been learned through our investigations above.

REFERENCES

B. Ahmed and J.G. Robinson, "Estimates of the foreign-born population: 1980–1990," *Technical working paper No. 9*. Population Division, U.S. Census Bureau, Washington, D.C., 1994.

J. Berger and C. Cunningham, *Bison: Mating and Conservation in Small Populations*, Columbia University Press, New York, 1994.

M.L. Bonnell and R. K. Selander, "Elephant seals: Genetic variation and near extinction," *Science*, 184 (1974) 908–9.

E. Callenback, *Bring back the buffalo! A sustainable future for America's Great Plains*, Island Press, Washington, D.C., 1996.

M. Eddy, "White buffalo causes stir," *Denver Post*, 13 January, 1996.

S. Foss, "White Cloud keeps her distance," *Grand Forks Herald*, 8 December, 1996.

P.A. Fuerst and T. Maruyama, "Considerations in the conservation of alleles and genic heterozygosity in small managed populations," *Zoo Biology*, 5 (1986) 171–79.

M.S. Garretson, *The American Bison*, New York Zoological Society, New York, 1938.

V. Geist, *Buffalo Nation: History and Legend of the North American Bison*, Voyageur Press, Inc., Stillwater, Minnesota, 1996.

D.L. Hartl, *Principles of Population Genetics*, Sinauer, Sunderland, Massachusetts, 1980.

B. Hodgson, "Buffalo: Back home on the range," *National Geographic*, 5 (1994) 64–89.

W.T. Hornaday, "The extermination of the American bison, with a sketch of its discovery and life history," in *The Annual Report (1889) of the Smithsonian Institute*, 367–548.

W.S. Klug and M. R. Cummings, *Concepts of Genetics*, Prentice-Hall, Upper Saddle River, New Jersey, 1997.

G.T. Kurian, *Datapedia of the United States 1790-2000*, Bernan Press, Lanham, Maryland, 1994.

A. Matthews, "The Popper's and the plains," *New York Times Magazine*, 24 June 1990.

A. Matthews, *Where the Buffalo Roam: The Storm Over the Revolutionary Plan to Restore America's Great Plains*, Grove Press, New York, 1992.

J.N. McDonald, *North American Bison: Their Classification and Evolution*, University of California Press, 1981.

T. McHugh, *The Time of the Buffalo*, Alfred A. Knopf, New York, 1972.

M. Meagher, "Bison bison," *Mammalian Species*, 266 (1986) 1–8.

E. Nevo, A. Beiles, and R. Ben-Shlomo, "The evolutionary significance of genetic diversity: Ecological, demographic, and life history correlates," in *Evolutionary Dynamics of Genetic Diversity*, G. S. Mani, ed., Springer-Verlag, New York, 1984.

S.J. O'Brien, D. E. Wildt, D. Goldman, C. R. Merril, and M. Bush, "The cheetah is depauperate in genetic variation," *Science*, 221 (1983) 459–462.

J.A. Paulos, *Innumeracy: Mathematical Illiteracy and its Consequences*, Hill and Wang, New York, 1988.

P.H. Raven, L.R. Berg, and G. B. Johnson, *Environment*, Saunders College Publishing, 1995.

F.G. Roe, *The North American Buffalo*, Second ed., University of Toronto Press, Toronto, 1970.

P.G. Russell, *Genetics*, Second ed., Scott, Foresman, and Co., Glenview, Illinois, 1990.

R. Wronski, "White buffalo fulfills a tribal prophecy," *Chicago Tribune*, 11 September, 1994.

Introduction to the Mathematics of Populations, Invasions, and Infections[1]

Martin E. Walter
University of Colorado

INTRODUCTION

The human population of the world is growing by about a billion persons per decade. The human population of the United States is growing between 2 to 3 million persons per year. It is very important to understand what this growth means for your life, for there will be substantial effects—not all of them good.

A commercial airline pilot once said to me that she violates the law of gravity all the time. What I was not quick enough to reply at the time was: "Only as long as there is fuel in your tanks." Humans can create *niches*, i.e., functional places in the ecosystem or "jobs," for themselves in two ways: by taking niches from other living organisms or by creating *temporary* niches beyond an ecosystem's carrying capacity by drawing down resources faster than they are being replaced. This latter situation is called *overshoot*. It has been persuasively argued by William R. Catton, Jr.[2] that today's industrial civilization is in a state of overshoot created by technology, which makes extra human niches by drawing down reserves of fossil fuels and other resources. If Catton is right, then we had better try to land our civilization softly while there is still fuel in the tank. The alternative appears to be an out-of-human-control crash.

---

[1] Dedicated to the Memory of David Brower.
[2] See [4].

1   A SHORT HISTORY OF HUMAN POPULATION SIZE

When studying something in the real world it is a good idea to get some actual facts whenever possible. An interesting, scholarly book by Joel E. Cohen[3] deals with many aspects of human population growth. In particular, he gives a short history of the size of the human population.[4] At 10,000 B.C. Cohen estimates the earth's human population was between 2 million and 20 million people. At A.D. 1 he estimates the world population to have been between 170 million and 330 million people, i.e., roughly a quarter billion people.

Between A.D. 1600 to A.D. 1650 the earth's human population reached about a half-billion. Between 1800 and 1850 the population reached 1 billion. In about 1930 the population hit 2 billion. In 1960 the population total was 3 billion, an increase of a billion in just 30 years. Fourteen years later, in 1974, the population reached 4 billion. Twelve years later, in about 1986, the population passed the 5 billion mark. As I write this in 1997, estimates of the world's human population in 1996 are about 5.8 billion.

Exercise 1.1.

(a) On a sheet of paper draw a pair of coordinate axes, i.e., a horizontal line representing time and a vertical line representing population. Now graph the earth's total human population versus time, from about 1 A.D. until the present. Remember that a billion is one thousand times a million.

(b) What is the world's total human population at the time you read this? Where can you find this information? How reliable, or how accurate, is this information?

(c) Estimate the doubling time, i.e., the time it took the human population to double, at least four different times during the last 12,000 years. Do you notice a trend?

The data that we have just discussed (and that you just graphed) does not precisely fit any simple mathematical model. There is no simple formula (or formulas) that "explain" the data above. Nevertheless, there are some simple, meaningful models that can be created. First, let us look at the above numbers from the point of view of an ecologist.

2   A FUNDAMENTAL AXIOM OF POPULATION ECOLOGY

Ecologists define the *niche* of a living organism to be that organism's functional address in the environmental system in which it lives. Intuitively, we can think of a niche as a job. Thus a grizzly bear in Denali Park, Alaska, occupies the niche of a grizzly bear and does the job of such a bear. To understand this niche we have to know all of the bear's activities and how these activities affect the bear's environment. A niche is not just a physical place that we can locate with three space coordinates, varying in time—a multitude of other dimensions are involved. How does the bear relate to, interact with, and/or affect other species and other physical parts of its environment? What, where and how does it ingest?

---

[3]See [5].
[4]See [5, Chapter 5].

What, where and how does it excrete? Does the bear interact with wolves? The list of questions is truly endless, and probably not all of them can be answered.

Ecologists[5] have observed the following to be true of any species capable of increasing its numbers, i.e., all existing normal species.[6]

Ecological Axiom on the Size of a Population. *The eventual number of individuals of a given species, i.e., the eventual size of the population of that species, is determined by the number of niches available for occupation by that species (not by the rate at which that species reproduces).*

Any given species will fill up all the niches available to it. Apparently humans are not different—so far—from any other species in this regard. This allows us to view the population data for humans in Part 1 above in a different light. The graph of the human population in Exercise 1.1 does not just depict the growth in human numbers. This graph also shows approximately the number of niches available for human occupation at various times—niches that humans quickly occupied. Humans have used technology, including agriculture, to increase the number of niches for humans on earth.

Exercise 2.1.

(a) Look anew at your graph in Exercise 1.1(a). Discuss what technological innovations occurred at various times to increase the number of niches available to humans. For example, when was agriculture first invented? When were fossil fuels used to mechanize agriculture and virtually everything else?

(b) In Royal Chitwan National Park, Nepal, and in Matopos National Park, Zimbabwe, armed troops protect wildlife from poachers and other excessive human intrusions. Discuss how this situation relates to our Ecological Axiom above.

3   A SIMPLE MATHEMATICAL MODEL OF UNCHECKED POPULATION GROWTH

The English clergyman-scholar, Thomas Robert Malthus, states in his 1798 work, *Essay on the Principle of Population*, "Population, when unchecked, increases in a geometric ratio. Subsistence increases only in an arithmetic ratio."

William R. Catton[7] restates this proposition of Malthus as: "The cumulative biotic potential of the human species exceeds the carrying capacity[8] of its habitat." Catton goes

---

[5]See [6, Chapter 2].

[6]There are abnormal examples of species suffering toxic effects of pollution, for example, that become sterile or nearly sterile. These species become extinct unless something is done to reverse the situation. In this situation, one can interpret such a polluted environment as offering few or no niches for the given species—hence the following axiom could still be seen as valid even in this case.

[7]See [4, Chapter 8].

[8]The carrying capacity of an ecosystem for a given species is defined to be the maximum population of that species that can be supported indefinitely by that ecosystem.

on to generalize this proposition by replacing the words "of the human species" by the words "of any species." Catton says that the phrase "biotic potential" refers to the total number of offspring a parental pair would be theoretically capable of producing.

Malthus has been roundly denounced by some authors as being flat-out wrong. The graph from Part 1 is often offered as proof. Catton's thesis is that many of the niches now available for humans were created (and are being maintained) by the extraction of nonrenewable resources, such as coal, oil and natural gas. Through technology humans have made it appear that the carrying capacity of the earth for humans is larger than the number of humans. When these resources are gone, unless replacements are found, the true carrying capacity of the earth for humans will be much smaller than the number of humans and large die-off will follow.

Whether or not you agree with Catton, I contend that we can decide if Malthus was right about unchecked populations growing geometrically. To do this I will use a device which I call a *Schwartz Chart*.[9] What follows is an example of a *mathematical model* of a population.

Let us assume that we are dealing with a population of a country that roughly satisfies the six assumptions below. This first model has some obviously "wrong" assumptions, but our first model is not totally unrealistic either. For example, it models approximately the population of Costa Rica a quarter century ago. While we are constructing this first model ask yourself how we might make the model more accurately reflect a "real" population. You will get a chance to implement your more realistic assumptions in Part 5. For the moment, let us pretend that there is a country out there in "Mathland" that satisfies the following six over-simplified assumptions.

Six Population Rules in Mathland

1. Everyone in the population falls into one of the following four "compartments" or "boxes" according to age: Children (people 15 years of age or less but greater than 0), Reproducers (people aged 30 years or less but greater than 15), Parents (people aged 45 years or less but greater than 30), Grandparents (people aged 60 years or less but greater than 45).

2. Everyone dies at age 60, not before, not later.

3. Exactly half of the people in each "age compartment" are male, half are female.

4. Everyone is monogamous, no one is celibate: while a person is a Reproducer he/she is paired (by marriage) with exactly one partner of the opposite sex.

5. Each couple of Reproducers has exactly 4 Children. While a person is in a compartment other than the Reproducer compartment he/she has no Children.

6. We partition time into intervals 15 years long. The first interval is 15 years or less, but more than 0 years. The second interval is 30 years or less but greater than 15 years, and so on. At the end of the first 15 year interval we assume that we have 10 Children and 0 people in each of the other 3 compartments.

---

[9]See [9].

Following the above six rules or assumptions we can fill out the Schwartz Chart below in Table 1.

TABLE 1

At the end of year	Children	Reproducers	Parents	Grandparents	Total Pop
15	10	0	0	0	10
30	20	10	0	0	30
45	40	20	10	0	70
60	80	40	20	10	150
75	160	80	40	20	300
90	320	160	80	40	600
105	640	320	160	80	1200
120	1280	640	320	160	2400
135	2560	1280	640	320	4800
150	5120	2560	1280	640	9600
165	10240	5120	2560	1280	19200
180	20480	10240	5120	2560	38400
195	40960	20480	10240	5120	76800
210	81920	40960	20480	10240	153600
225	163840	81920	40960	20480	307200
240	327680	163840	81920	40960	614400
255	655360	327680	163840	81920	1228800
270	1310720	655360	327680	163840	2457600
285	2621440	1310720	655360	327680	4915200
300	5242880	2621440	1310720	655360	9830400

In the following exercise we test to see if you understand how we got Table 1.

Exercise 3.1.

(a) In the second interval of time, 30 years or less but greater than 15 years, how many Reproducers are there? Why?

(b) In the second row, i.e., second interval of time, how many married Reproducer couples are there? How do you get this number?

(c) How many Children are there in the second interval of time? How do you get this number?

(d) How many Parents and Grandparents are there in this second interval of time? Why?

(e) What is the total population in the second interval of time, i.e., in the second row of our Schwartz Chart?

(f) Carefully calculate the entry in each compartment in the third row of the Schwartz Chart. Do your answers agree with those in Table 1?

(g) Fill out at least the next six rows of the Schwartz Chart, not forgetting to calculate the total population in each row, i.e., each interval of time.

I now want to focus attention on two questions. Question 1: How many years does it take the population in Table 1 to double?[10] Question 2: Does this doubling time depend on the number of Children that we start with in the first row?

In Table 1, our first question is easy to answer if we look at row four and beyond. The total population in row four is 150. The total population in the next, or fifth, row is 300—exactly twice the total population in the fourth row. You can check to see that the following rule holds for rows beyond the fourth: *The total population in a row is two times the total population in the previous row.*

This is precisely what Malthus meant by "geometric growth" of a population. In our simple model, the total population in a given row (or time interval or "generation") is a fixed multiple (in this first case the multiple is 2) of the total population in the previous row (or time interval or "generation").

Thus the answer to the question: "How long does it take for this Schwartz Chart population to double?" is 15 years. This coincides (in this case) with the time it takes to go from one row to the next, or one "generation" to the next.

To answer the second question, do the next exercise.

Exercise 3.2.

(a) Replace the number of Children in the first row by 4, i.e., modify assumption (6). Recalculate the Schwartz Chart with all of the other axioms unchanged. What doubling time do you get now?

(b) Replace the number of Children in the first row by 20. Recalculate the Schwartz Chart leaving all of the other axioms unchanged. What doubling time do you get now?

(c) Are you willing to make a guess as to the answer of Question 2: Is the doubling time affected by changing the initial number of Children? What do you need to do to "prove" your answer?

Now let's tackle a model that is a little bit harder. Let's go back to the "six population rules in Mathland" and change rule (5). Instead of having each Reproducer couple produce 4 Children, let each Reproducer couple have 3 Children. Now consider the following exercise.

Exercise 3.3.

(a) Construct a Schwartz Chart for a population wherein each Reproducer couple has 3 Children, and all other previous rules are unchanged. Again, start the first row of your chart with 10 Children and 0 persons in each of the other compartments. (Don't worry if you get fractional people, that's OK in Mathland! If you want to avoid fractional

---

[10]The time it takes a population to double is called the *doubling time* of the population.

people for a few rows start in the first row with a number of children that is divisible many times by 2, for example, 32 or 256.)

(b) Although it is not so easy as in the 4 Children per Reproducer couple case, estimate the time it takes for the population in (a) (of this problem) to double.

(c) If each Reproducer couple has 3 Children is the doubling time longer or shorter than when each Reproducer couple has 4 Children?

(d) After the fourth row of the Schwartz Chart in (a) can you come up with a rule for the growth of the total population? Can you fill in the blank in the following? After the fourth row, the total population in a row is ___ times the total population in the previous row.

(e) Can you guess a relationship between the number in the blank in (d) and the doubling time in (b) (of this problem)? This question is tough.

4 A DEEPER ANALYSIS OF SCHWARTZ CHARTS

If you understood everything (or almost everything) so far, you know enough to do some interesting modeling problems. Realistically, it will be an immense help if you have already studied logarithms. However, there is quite a bit you can do even if you have never seen logarithms—or you have totally forgotten what you thought you once knew about them. (If you prefer, you can safely skip now to Part 5.)

If you have access to a calculator it is quite likely that there is a button on that calculator labeled *log* or LOG. Let me remind you of the definition of the *log* as it appears on your calculator.

Definition of *log*. *The logarithm of a (positive) number x to base ten, written $\log_{10} x$ or $\log x$ or LOG x, is the power to which ten must be raised to get x.*

If you have seen logarithms before, this exercise should be fairly easy. If you have not, this exercise may be pretty challenging—but try it anyway!

Exercise 4.1.

(a) What are $\log 10$, $\log 100$, $\log 1000$, $\log .1$, $\log .01$? Find these numbers on your calculator. You should get the numbers $1, 2, 3, -1, -2$, respectively.

(b) What is 10^2? What is $10^{\log 100}$? Answer: 100 in both cases. This problem is supposed to remind you what it means to raise 10, say, to the power 2. It is also supposed to remind you that the definition of logarithm of x to base 10 means that $10^{\log x} = x$ for any positive number x.

(c) What is $\log 2$? You should get a number that agrees with .3010299957 at least up to the last digit or however many digits your calculator gives.

(d) What approximately is $10^{.3010299957}$?

(e) What is $\log 3$? Answer to 10 digits: .4771212547.

(f) What is log 6? Answer to 10 digits: .7781512504. This exercise is supposed to remind you of the fact that if x and y are two positive numbers then $\log xy = \log x + \log y$. Is $\log 6 = \log 3 + \log 2$?

(g) What is log 9? Answer to 10 digits: .9542425094. This exercise is supposed to remind you that $\log x^r = r \log x$ if x is a positive number and r is a number. Is $\log 9 = 2 \log 3$?

Logarithms can do "magic." Before we can see a picture of the magic, however, we need a graph. Go back to Table 1 and graph the total population versus time for time between, say, 15 and 150 years. Now compare your graph to Figure 1.

Total Population vs. Time
(from Table 1)

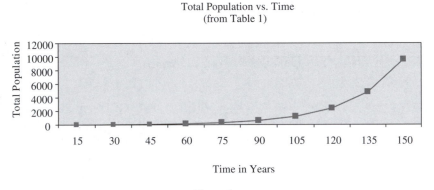

Figure 1.

The graph in Figure 1 is the result of plotting the total population versus time in the Schwartz Chart wherein each Reproducer couple has 4 Children.

Now use the log button on your calculator. On the right side of Table 1, add another column and title it "Log Total Population." This is Table 2 (below). Thus to the right of "total population" you will have written "log of total population" at the head of a seventh column. Now, just to the right of the 10 in the sixth column write the number, which is log 10, which you find using your head or your calculator. If you do this on your calculator you should get 1.000000000 up to 10 digits. Continue in this way filling out the seventh column. For example, the next entry should be log 30 = 1.477121255 up to 10 digits. Once you have filled out the seventh column of "log of total population," at least for time up to 150 years, graph the "log of total population" column of numbers versus time. Once you have done this compare your graph to the one below in Figure 2.

What we have just done is apply the log operation to the curved graph in Figure 1, and the result we get is the straight line graph in Figure 2. (Well, the graph in Figure 2 is almost a straight line. At least after the time of 45 years the graph is indeed a straight line.)

Thus sometimes a curved line becomes a straight line after applying the log operation. Logarithms have made things simpler!

Now we calculate the slope of the line in Figure 2 (for time greater than 45 years). It turns out that this slope is very intimately connected with the doubling time of the popula-

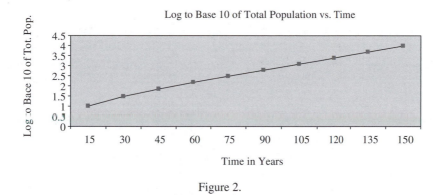

Figure 2.

tion in the Schwartz Chart. In fact, if I know this slope I can immediately write down the doubling time. But first, let's calculate the slope.

Since the graph in Figure 2 is a straight line for time greater than 45 years, all we need do is use two points on the graph to calculate the slope. We pick (whimsically) $t = 75$ and $t = 135$ as two times on the horizontal axis. The coordinates of the point on the graph just above time equal to 135 years are $(135, \log 4800)$. To see why this is, go back to Table 1

TABLE 2

At the end of year	Children	Reproducers	Parents	Grandparents	Total Population	Log Total Population
15	10	0	0	0	10	1
30	20	10	0	0	30	1.477121255
45	40	20	10	0	70	1.84509804
60	80	40	20	10	150	2.176091259
75	160	80	40	20	300	2.477121255
90	320	160	80	40	600	2.77815125
105	640	320	160	80	1200	3.079181246
120	1280	640	320	160	2400	3.380211242
135	2560	1280	640	320	4800	3.681241237
150	5120	2560	1280	640	9600	3.982271233
165	10240	5120	2560	1280	19200	4.283301229
180	20480	10240	5120	2560	38400	4.584331224
195	40960	20480	10240	5120	76800	4.88536122
210	81920	40960	20480	10240	153600	5.186391216
225	163840	81920	40960	20480	307200	5.487421211
240	327680	163840	81920	40960	614400	5.788451207
255	655360	327680	163840	81920	1228800	6.089481203
270	1310720	655360	327680	163840	2457600	6.390511198
285	2621440	1310720	655360	327680	4915200	6.691541194
300	5242880	2621440	1310720	655360	9830400	6.99257119

and notice that at the end of 135 years the total population is 4800, thus $(135, \log 4800) = (135, 3.681241237)$. The other point is $(75, \log 300) = (75, 2.477121255)$.

Now to get the slope of the line through these two points we take the difference in the second coordinates and divide by the difference (in the same order) of the first coordinates, and we get

$$\frac{3.681241237 - 2.477121255}{135 - 75} = \frac{1.204119982}{60} = .0200686663.$$

One way to check if we really have a straight line in Figure 2 for time greater than 45 years is to take two other points, corresponding to times greater than 45 years, and see if we get the same answer for the slope. This is the content of the next exercise.

Exercise 4.2.

(a) Pick two other points corresponding to times greater than 45 years in the graph in Figure 2 and calculate the slope of the line between them. Compare your answer with the answer .0200686663 that we just obtained above. How close are they?

(b) What do you need to do to prove that the graph in Figure 2 for time greater than 45 years is indeed a straight line?

Now if we take $\log 2$ and divide it by 15 we get

$$\frac{\log 2}{15} = .0200686663.$$

The following formula (although not proved here) is true, and it is the reason why the slope above is equal to $\log 2/15$. The doubling time, written t_d, for the population in a Schwartz Chart that satisfies the previous six population rules except that in rule (5) each Reproducer couple has exactly K Children is given by the following formula:

$$t_d = \frac{15 \log 2}{\log \frac{K}{2}}.$$

We could rewrite this last formula in the form

$$t_d = \frac{\log 2}{\frac{\log \frac{K}{2}}{15}} = \frac{\log 2}{\text{slope of "log total population vs time"}}.$$

Let's use this formula to check up on some of our earlier work.

Exercise 4.3.

(a) Using the formula for t_d above, calculate the doubling time of a Schwartz Chart population if $K = 4$, i.e., each Reproducer couple has 4 Children. Do you get the same answer as we obtained before?

(b) Using the formula above for t_d, calculate the doubling time of a Schwartz Chart population if $K = 3$, i.e., each Reproducer couple has 3 children. How does this compare to your answer to Exercise 3.3(b)?

(c) Can you answer Exercise 3.3(e) now?

(d) Recapitulate, for the case of each Reproducer couple having 3 Children, the calculation of " slope of log total population versus time" (which we did above in the case of each Reproducer couple having 4 Children). Take this value for the slope and divide it into $\log 2$ and see if you get the same answer for t_d as you did in (b) of this problem.

The following exercise is tough if you are not familiar with logarithms. If you are, you might enjoy the challenge.

Exercise 4.4. Prove the formula $t_d = (15 \log 2 / \log \frac{K}{2})$ stated above. Hint: First derive a formula for the total population as a function of time, denote this formula by $P(t)$. Note that $P(t + t_d) = 2P(t)$. Then use the properties of logarithms reviewed in Exercise 4.1. to take the log of this formula. It turns out that $\log P(t)$ has a graph that is a straight line (for time greater than 45 years). You can find the answer to this problem in [10].

5 SOME MATHEMATICAL PROPERTIES OF POPULATIONS

I have a confession to make. Although when I first started using Schwartz Charts in my class I used a sheet of paper and my calculator to fill them out, I soon started using a larger computer (like a Macintosh) and a computer program called a *spreadsheet*. The particular program I use is called EXCEL, but that does not matter. This section can be done with a sheet of paper and a calculator (or even without a calculator), but things move a lot faster if you have access to a computer with a spreadsheet program. Someone can help you at your local school or library.[11]

The following exercises explore some (but not all) variations of Schwartz Charts and we discover some interesting mathematical facts that probably say more than we realize about the "real world."

Exercise 5.1. If we go back to the six population rules and "replace the 15 with a 20" throughout we will get a Schwartz Chart that more closely resembles the United States. Thus everyone dies at 80 and all of our compartments and partitions are based on 20-year intervals instead of 15-year intervals. How does replacing the 15 with a 20 affect doubling times? Can you give a formula in this case for t_d?

Exercise 5.2. You might find the answer to this problem surprising—or counterintuitive. Suppose you have two Schwartz Charts with most of our six original assumptions unchanged: 15-year intervals for compartments and partitions of time, everyone dies at 60. In one Schwartz Chart every Reproducer couple has 3 Children and no other compartment

---

[11]Librarians tend to be overworked these days; try getting help from a classmate first.

has any Children. In the second Schwartz Chart each Reproducer couple has 1 Child and each Parent couple has 3 Children. Compare the doubling times of these two populations.

Many people have pondered how we might stabilize the world's human population. The next problem challenges you to find a way to stabilize, at least mathematically, a "Schwartz Chart population."

Exercise 5.3. Play the following game. By making various assumptions, see if you can arrange a "Schwartz Chart like" population that is stable. By this I mean a population that neither doubles endlessly nor goes extinct.[12] Can you arrange it so that the population remains below some predetermined upper limit but never goes to zero? A simple model that achieves this is our Schwartz Chart with each Reproducer couple having 2 Children. If you do not allow this example, what else can you come up with? Do you think that any real population will satisfy your assumptions?

Exercise 5.4. Modify your Schwartz Chart to make it as realistic as possible. For example, introduce a death rate in each compartment. Introduce the possibility of birth rates in more than one compartment. How would you take immigration and emigration into account?

Exercise 5.5. An important property of any population is its *age distribution*. For example, if over half the population of the United States were over 65, the implications for Social Security and medical care would be severe. If over half of a population were under 15, then the implications for the educational system would be severe. Pick a Schwartz Chart. Now for a given row, plot on a bar graph the number of people in each of the compartments: Children, Reproducers, Parents and Grandparents. What is your graph like, and how does it change from one row ("generation") to the next? Change your assumptions about your Schwartz Chart and see what affect it has on the age distribution. For example, how does family size affect age distribution?

Exercise 5.6. In this exercise we study a concept called *population momentum*. If a country decides that its population is growing too large and wants to reduce it by passing some law limiting family size (as in China, for example), it will take some time for the law to have an effect. Suppose you have a population that is growing quite fast, say a Schwartz Chart with each Reproducer couple having 4 Children. Pick a row, say the tenth row, and from that row onward reduce the number of Children per Reproducer couple to 1, leaving all the other assumptions the same. How long does it take for the total population to decrease to half of the total population in the tenth row?

I have not said anything explicit about the role that competition plays in controlling population growth. Much has been written about competition, and in fact I discuss it a little elsewhere.[13] Competition can occur between species, and it can occur between subsets of

---

[12] A population is extinct when it has 0 members in it.
[13] See [10].

humans. In this article I will leave the subject open for you to explore on your own. How would you build competition into a Schwartz Chart? From a pure mathematician's point of view some of the mathematical patterns encountered in studying competition will be presented in Part 7 when we look at epidemics/infections.

6 A QUICK LOOK AT INVASIONS

By definition when a life form from outside an ecosystem, i.e., an exotic, appears inside that ecosystem—an invasion has occurred. "The ancient barriers, the molds within which ecosystems formed, are crumbling. The resulting spread of exotic animals is now second only to habitat loss as a category of ecological destruction, according to Harvard biologist Edward Wilson, an authority on biodiversity. Vernon Heywood, a specialist on plant invasions, considers exotic plants among the most serious threats to natural plant communities."[14] I refer you elsewhere for detailed biological, economic, and historical data.[15] Suffice it to say that the consequences of the corresponding extinctions and ecological disruptions are large and warrant far more attention than humans have given the problem thus far. The economic impacts alone deserve much more attention than has been given to date.

I have one brief, simple yet important mathematical observation to make with regard to invasions. When an exotic species successfully invades and establishes itself in an ecosystem, the situation (at least initially) is usually characterized by the words of Malthus—"unchecked growth." Thus, in the beginning of a successful invasion, the simple mathematics of "geometric growth" that we encountered in our first Schwartz Charts can be expected to apply.

If I were to start constructing a more detailed model of some specific invasion I would start with the basic characteristics of a successful invader: (1) preference for disturbed habitat, (2) efficient dispersal, (3) rapid population growth, and (4) opportunistic feeding. The cheat grass invasion of the American west, zebra mussels' invasion of the Great Lakes and the invasion of just about everywhere by rats—these are all examples of "weedy" invasions.[16]

7 A SIMPLE MODEL OF AN EPIDEMIC

Although humanity is in the running for being the most successful invader of all time, Nature is not without tools in dealing with us. Although we have virtually eliminated species groups such as tigers, wolves, and certain trees, we have not made a dent in the collective biomass of insects, bacteria, viruses, and "sub" viruses. Infectious diseases, for example, are still formidable competitors, killing 16.5 million people worldwide in 1993. To put this in perspective, the worldwide human death toll from cancer in 1993 was 6.1 million; from

---

[14]See [2, Chapter 6, p. 96], which is an article entitled "Understanding the Threat of Bioinvasions" by Chris Bright.
[15]For example, see [2] and [7].
[16]See [2, p. 102].

heart disease was 5 million; from cerebrovascular diseases such as stroke was 4 million; and from respiratory diseases like bronchitis was 3 million.[17]

For an in-depth look at infectious disease, including relevant, interesting mathematics there is an encyclopedic reference.[18] We will concentrate on a simple, interesting and informative model of an epidemic which is quite well known.[19] Before going on it is worth noting that the population growth of humans observed in Part 1 makes it easier for infectious disease to find at least one of us and then spread to others of us.

The following classic disease model can be studied with just a pencil and paper, or with a pencil, paper and calculator, or with a spreadsheet as we discussed in Part 5.

When modeling epidemics, as in our population models, it is useful to partition the population into compartments or boxes; but now the compartments depend on the person's relationship to the disease being studied.

Imagine a population of humans exposed to an outbreak of influenza, as is the case each winter. The time it takes the disease to play itself out—from exposure, through incubation and full-blown illness, to cure (or death)—is relatively short compared to the natural generational cycles. In fact, the whole annual epidemic usually lasts less than the winter season. Thus we can ignore a lot of complicating factors in our model—and we will.

Six Influenza Rules in Mathland

1. We assume that the total size of the population affected is constant during our flu outbreak.

2. We assume that at any particular time every individual belongs to exactly one of the following compartments or "boxes:"

    (a) Susceptibles (those well persons who have not had the disease this season and have no immunity),

    (b) Infected (persons who have the flu in some stage),

    (c) Cured (those well persons who have had the disease, recovered and have short-term immunity),

    (d) Dead (those who die from the disease).

    In our simplified model we will ignore the existence and level of effectiveness of flu shots. We ignore also any deaths not due to the flu, and we ignore any births.[20] If t is the time in days (in our case this is an integer that also denotes a row in our spreadsheet), we let $S(t)$, $I(t)$, $C(t)$, and $D(t)$ denote the number of people in each of the respective compartments for the row t. Note that rule (1) above can now be written: $S(t) + I(t) + C(t) + D(t) = N$, where N is a constant.

---

[17] See [2, Chapter 7], entitled "Confronting Infectious Diseases," by Anne E. Platt.

[18] See [1].

[19] See [8].

[20] At the end of Part 7 see Figure 3 for the complete spreadsheet data from one influenza model. On our spreadsheet (either a piece of paper or a computer screen) we have our first column labeled "time," measured in days; the next column, "Susceptible;" the third column, "Infected;" the fourth column, "Cured;" and the last column; "Dead."

3. We will assume that there is a number (which does not depend on time), R_i, called the *infection rate*, which satisfies:[21]

$$S(t+1) - S(t) = -R_i S(t) I(t).$$

4. We will assume that there is a number (which does not depend on time), R_c, the *cure rate*, which satisfies:[22]

$$C(t+1) - C(t) = R_c I(t).$$

5. We will assume that there is a number (which does not depend on time), R_d, the *death rate*, which satisfies:

$$D(t+1) - D(t) = R_d I(t).$$

6. The rates of infection, cure and death also satisfy:

$$I(t+1) - I(t) = R_i S(t) I(t) - R_c I(t) - R_d I(t).$$

Exercise 7.1.

(a) Suppose you have m red dots and n blue dots. How many line segments can you draw that begin at a red dot and end at a blue dot? Answer: mn, or m times n. Why?

(b) If m represents the number of Infected and n represents the number of Susceptible, then mn represents roughly the number of "contacts" that can occur between these two groups. A certain fraction of these contacts will result in transmission of disease from the Infected to the Susceptible. If $R_i = .00001$, then we are assuming that that fraction is "one in one hundred thousand." (This is just an educated guess on my part. After you have read and understood the model we are constructing, feel free to experiment and see how the model responds to changes in the infection rate. The consequences are not dire since this is Mathland.) Thus the infection rate tells us how virulent a particular flu is. Do you think this is a reasonable value for the infection rate?

(c) We can rewrite rule 3 as $S(t+1) = S(t) - R_i S(t) I(t)$. Do you see that this equation says that the number of Susceptibles in one row is equal to the number of Susceptibles in the previous row less the number of Susceptibles that have moved over to the Infected box?

Exercise 7.2.

(a) The equation $C(t+1) - C(t) = R_c I(t)$ says that the change in the number of people in the Cure box, in a given row (that is, in a given day), is a certain fraction of the persons in the Infected box. If $R_c = .7$, what fraction of Infected become Cured each day? Is this a reasonable fraction, based on your experience with the flu?

---

[21] See Exercise 7.1. for some justification for this assumption.
[22] See Exercise 7.2. for a short discussion of this assumption and the other two assumptions (rules) that follow.

(b) The equation in part (a) can be rewritten as $C(t+1) = C(t) + R_c I(t)$. Do you see what this says about the flow into the Cured box from the compartments (or boxes) from the previous row?

(c) The equation $I(t+1) - I(t) = R_i S(t) I(t) - R_c I(t) - R_d I(t)$ tells us that the change in the number of people in the Infected box is given as a sum; namely, the number of people who move from the Susceptible box to the Infected box, plus (minus) the number of people who move from the Infected box to the Cured box, plus (minus) the number of people who move from the Infected box to the Death box. Explain each of the minus signs in this equation and the minus sign in the equation in rule 3.

We can rewrite the equation as

$$I(t+1) = I(t) + R_i S(t) I(t) - R_c I(t) - R_d I(t).$$

Can you explain in words what this equation says?

(d) Does it follow from the equations in rules 3, 4, 5, and 6 that

$$S(t) + I(t) + C(t) + D(t) = S(t+1) + I(t+1) + C(t+1) + D(t+1)?$$

(e) My educated guesses for the values of the cure rate and the death rate are: $R_c = .7$ and $R_d = .001$, respectively. Do you think that these numbers are "reasonable?" You should feel free to experiment with these values as you were previously encouraged to experiment with the infection rate.

(f) If the cure rate and the death rate were to add up to 1, what would this say about our epidemic model?

Following, we have the complete spreadsheet output from a model obeying the six rules above, with the values I have indicated for the infection, cure and death rates. Also, I assumed that the total population, N, was 100,000. The computer rounded off all the numerical entries to integers to make the output look better (and so it would all fit easily on one page). On the page following the numerical data is a graph showing the rise and fall of the infected persons, and the death toll. I did not create this model to fit real data from any specific epidemic I have studied. However, see Part 8.

TABLE 3

Time	Susceptible	Infected	Cured	Dead	Time	Susceptible	Infected	Cured	Dead
0	100000	1	0	0	39	63959	5306	30692	438
1	99999	1	1	0	40	60565	4980	34406	492
2	99998	2	2	0	41	57549	4505	37892	541
3	99996	2	3	0	42	54956	3940	41046	586
4	99994	3	4	0	43	52791	3343	43804	626
5	99991	4	6	0	44	51028	2765	46144	659
6	99987	5	9	0	45	49616	2237	48080	687
7	99982	6	12	0	46	48506	1779	49646	709
8	99976	8	17	0	47	47634	1395	50891	727
9	99968	11	22	0	48	46978	1082	51867	741
10	99958	14	30	0	49	46470	831	52624	752
11	99944	18	39	1	50	46084	635	53206	760
12	99926	23	52	1	51	45791	482	53651	766
13	99903	30	68	1	52	45570	365	53989	771
14	99873	39	89	1	53	45404	276	54244	775
15	99834	50	116	2	54	45278	208	54437	778
16	99784	65	151	2	55	45185	156	54582	780
17	99719	85	197	3	56	45114	117	54692	781
18	99634	110	256	4	57	45061	88	54774	782
19	99525	142	333	5	58	45022	66	54835	783
20	99383	184	433	6	59	44992	49	54881	784
21	99920	238	562	8	60	44970	37	54916	785
22	98964	308	729	10	61	44953	28	54942	785
23	98659	396	944	13	62	44941	21	54961	785
24	98268	510	1221	17	63	44931	16	54976	785
25	97768	653	1578	23	64	44924	12	54987	786
26	97129	834	2035	29	65	44919	9	54995	786
27	96319	1059	2619	37	66	44915	6	55001	786
28	95299	1337	3360	48	67	44912	5	55005	786
29	94025	1674	4296	61	68	44910	4	55009	786
30	92452	2074	5468	78	69	44908	3	55011	786
31	90534	2538	6919	99	70	44907	2	55013	786
32	88237	3056	8696	124	71	44906	2	55015	786
33	85540	3610	10835	155	72	44906	1	55016	786
34	82452	4168	13362	191	73	44905	1	55016	786
35	79015	4683	16280	233	74	44905	1	55017	786
36	75315	5100	19558	279	75	44904	0	55017	786
37	71474	5366	23128	330	76	44904	0	55018	786
38	67638	5440	26884	384	77	44904	0	55018	786

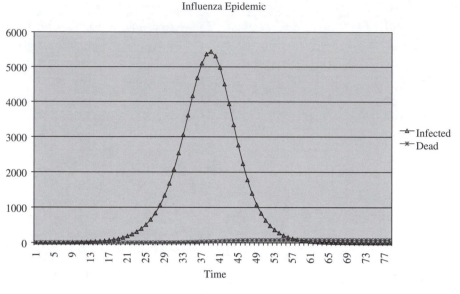

Figure 3.

8   IS THE AIDS EPIDEMIC REALLY CUBIC?

Recently it was asserted that the AIDS epidemic was not growing "exponentially" or "geo-metrically" but "cubicly." What does this mean and could it be so? Consider the following data.[23]

My colleague, Richard Holley, used *Mathematica* to see if a cubic polynomial would fit this data. Counting 1980 as $t = 0$, 1981 as $t = 1$ and so on, he came up with the following numerical result:

$$f[t] = .1197802197802154 - .04999229786380764t + .0000802345229581098t^2$$

$$+ .002706662962080952t^3.$$

Exercise 8.1.

(a) By hand, calculator, or computer, check a few of the data points from Figure 4 and see if this cubic polynomial does indeed fit the data.

(b) Does the curved line in Figure 4 become straight if you apply log to it?

When we plotted the graph of this cubic polynomial, the graph we got fit the data "perfectly." Richard Holley and another colleague of mine, Marion Dragonette asked our-selves if we could come up with a very simple model of an epidemic that would have cubic growth. We came up with one which we believe is too simple to capture all that is known about AIDS, but we nevertheless present it below in outline form as an exercise.

---

[23]See [3, p. 93].

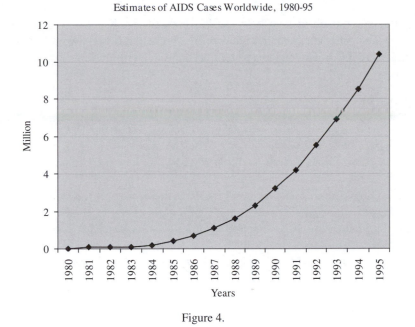

Figure 4.

Exercise 8.2. The following are rules for a very simple AIDS model. As you work on this model ask yourself how it might be changed to more accurately reflect what is known about AIDS.

RULE 1: Partition the population into 101 compartments, C_n, $n = 0, 1, \ldots, 100$. Assume that each person in C_n has n sexual partners a year.

RULE 2: The number of persons in C_n is K/n^3, where K is a constant independent of n and $n \neq 0$. For example, you can try $K = 10^8$. How do you want to handle the number of celibates?

RULE 3: A person from C_n only selects sexual partners from C_{n-1}, C_n, or C_{n+1}, with the obvious exceptions for C_0 and C_{100}. (You cannot have less than 0 partners, and we are arbitrarily not allowing more than 100 partners.)

RULE 4: Assume the epidemic starts with one infected individual in C_{100}.

If you have not had much probability theory, hopefully the following observations will be helpful. If R is the probability of becoming infected after an "encounter," then

$$\left[1 - R\left(\frac{\text{total number of infected people in } C_{n-1}, C_n, C_{n+1}}{\text{total number of people in } C_{n-1}, C_n, C_{n+1}}\right)\right]^n$$

is the probability of not getting infected after n partners. Subtracting this number from 1 gives the probability of getting infected after n partners.

I am tempted to attempt the construction of models in economics, politics, history and more! But I will have to wait. That does not mean that you have to wait! I hope that you have had some fun looking at the models in this article. I also hope you will have fun making up some models of your own. Lastly I hope that mathematical models will help all of us understand Nature and the "real world" a bit better. It can always be hoped that a better understanding might lead to a better life for all living things on earth.

REFERENCES

1. Roy M. Anderson and Robert M. May, *Infectious Diseases of Humans*, Oxford University Press, Oxford, New York, Tokyo, 1992.

2. Lester R. Brown, et al., *State of the World 1996*, W. W. Norton & Company, New York and London, 1996.

3. Lester R. Brown, et al., *Vital Signs 1996*, W. W. Norton & Company, New York and London, 1996.

4. William R. Catton, *Overshoot: The Ecological Basis of Revolutionary Change*, University of Illinois Press, Urbana and Chicago, 1982.

5. Joel E. Cohen, *How Many People Can the Earth Support?*, W. W. Norton & Company, New York and London, 1995.

6. Paul A. Colinvaux, *Why Big Fierce Animals Are Rare*, Princeton University Press, Princeton, New Jersey, 1978.

7. Charles S. Elton, *The Ecology of Invasions by Animals and Plants*, Methuen and Co. Ltd, London, 1958.

8. Michael G. Henle, "Forget Not the Lowly Spreadsheet," *The College Mathematics Journal*, **26** No. 4, September 1995, 320–328.

9. Richard H. Schwartz, "A Simple Mathematical Model for Population Growth," *Journal of Environmental Education*, **12** No. 2, Winter 1980–81, 38–41.

10. Martin E. Walter, *Mathematics for the Environment*, textbook in preparation.

Boom! Mathematical Models for Population Growth

Mohammad Moazzam
Salisbury University, MD
Richard H. Schwartz
College of Staten Island, NY

INTRODUCTION

There has been a boom in world population recently, largely due to an increase in life expectancies related to improved standards of living and advances in sanitation and medical technology. While it took until about 1850 for the world's population to reach one billion people, currently human population grows by approximately a billion people every 10 to 12 years.

According to the Population Reference Bureau's *1997 World Population Data Sheet*, world population in mid-1997 was 5.84 billion people. It is growing extremely rapidly and is projected to reach 6.89 billion by 2010 and 8.04 billion by 2025. While it took all of human history to reach its current population, at current rates of growth, world population is projected to double in only 47 years, to over 11.5 billion people. There is currently a population increase of about 88 million people every year. This means that the world population increase every 3 years is almost equal to the entire present population of the United States!

Populations in developing countries are growing especially rapidly; many countries in Africa, Asia and Latin America have population-doubling times of less than 30 years. While population growth has been slower in the United States, our population doubled in the last 56 years, from 132 million in 1940 to 268 million in 1997. Immigration is a large part of this growth.

With the Earth's population problems and the environment in jeopardy, mathematics can play a key role in warning and possibly solving a significant number of these problems. One of the most important uses of mathematics is to predict and evaluate population growth through mathematical models and to offer some manageable way of handling population growth and reducing pollution.

As you work through this paper, you will get familiar with three mathematical models for predicting, evaluating, and generalizing population growth. You will become acquainted with some other factors that are indirectly related to population growth and are big threats to our environment. Finally you will see some possible solutions and suggestions for a safer and cleaner environment.

Related Global Problems

Many important current global problems, including hunger, resource depletion, energy shortages, pollution, and poverty, are related to these very rapid population increases. Even without the expected continued sharp increase in population, an estimated 20 million people, including over 8 million infants, die annually due to hunger and its effects, and there are almost daily reports about environmental threats, such as depletion of the ozone layer, destruction of tropical rain forests and other habitats, global warming, acid rain, soil depletion and erosion, and air and water pollution.

Many people believe that rapid population growth is the greatest problem that the world currently faces. They emphasize connections between population increases and hunger, resource depletion, pollution, and other current problems. A group, "Zero Population Growth" (ZPG), argues that only with a stabilized population will the world's people be able to have clean air and water, decent places to live, meaningful jobs, and good educations. There is another group, "Negative Population Growth" (NPG), which claims that populations are already too high and should be reduced. It is very important that students be aware of issues related to population growth.

Fertility Rate and Future Population

A key indicator of future population growth is the fertility rate, the average number of children a woman has during her childbearing years. The replacement-level fertility, the numbers of children per couple that will produce a constant population, is theoretically 2. Actually, it is about 2.1 in the more developed countries, and up to 2.5 in the less developed countries, because of early deaths. If a country's fertility rate exceeds its replacement-level fertility, its population will not reach a constant value. A country with a fertility rate below its replacement fertility level may continue to grow in population for a while due to a momentum factor caused by a large number of children who will be entering their reproductive years, but eventually its population will stabilize. This analysis omits consideration of immigration.

While there are many factors that affect a country's fertility rate, one of the most important is the size and type of family considered desirable in that particular culture. (Another, of course, is the degree to which family-planning aids are available and accepted in the culture.)

MATHEMATICAL MODELS

The factors that must be considered in studying population increase in the world or in a particular country are many and varied. They include

a. the age structure,

b. the male-female ratio,

c. what percent of females have children,

d. at what age women have children,

e. how many children women have.

Thus, the study of population growth is complex.

In this article we present three models that will help students gain insight into population growth issues. Important concepts related to population growth can be understood using only very basic mathematical operations. These models are:

1. Tree Diagram Model

2. Tabular Model

3. Matrix Form Model

1   TREE DIAGRAM MODEL

Tree diagrams can help in analyzing a variety of problems in mathematics and science, particularly those related to probability. The mathematical ideas of tree diagrams and infinite series can give insight into how desired family size affects total fertility rate.

We will make the simplifying and reasonably accurate assumption that boy and girl children are equally likely. Note that with one child, there are two possible families (either a boy or a girl), and with two children, there are four possible families (BB, BG, GB, or GG). How many possible families are there for three children? four children? n children?

Let's consider the case of a family that has three children. We wish to know all possible families in terms of the sex of each child. Here is a tree diagram for this situation (B stands for boy and G for girl):

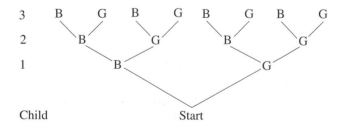

Figure 1. The diagram shows that there are eight different families, including BBG (boy–boy–girl), BGB, etc.

Suppose we want to know the probability that a family with three children will have exactly two girls. Since three of the eight families have two girls, the probability is 3/8.

Example 1   Consider a country in which all women marry and each couple wishes to have a son to provide economic help and security when the parents become old. Each woman will have children until she has one son, and then she will have no more children. What would the fertility rate be in such a country?

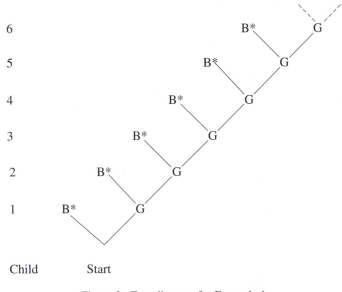

Figure 2. Tree diagram for Example 1.

Figure 2 shows a tree diagram for this case. B represents a boy, G a girl, and a * means that the desired family has been achieved, so that the woman will have no more children.

Analysis of the Results

The results can be analyzed as follows:

a. One-child families. After the first child, half of the couples will have a boy and half will have a girl. Hence, half of the couples (those having a boy) will stop having children after their first child.

b. Two-children families. Of the remaining couples (the half whose first child was a girl), half will have a boy for their second child. Hence, one quarter ($1/2 \times 1/2 = 1/4$) of the original couples will stop after having two children.

c. Three- or more-children families. As Figure 2 shows, this pattern continues, at every stage, half of the remaining couples will have a boy and thus stop having additional children.

Note that theoretically a woman could have a very large number of children before having a boy, but the probability of this happening is very small. For example, the probability of having 10 consecutive girls is $(1/2)^{10}$ or $1/1024$.

Average Number of Children per Woman

Based on our results so far, half of the women will stop at one child, a quarter of the women will stop at two children, an eighth of the women will stop at three children, and so on. The expected average number of children per woman will be:

$$\text{Average} = (1/2)(1) + (1/4)(2) + (1/8)(3) + (1/16)(4) + \cdots + (1/2^n)(n) + \cdots$$

You may wish to check that the nth term is $(1/2^n)(n)$.

Definition. *Any sum of numbers of the form*

$$S = a_1 + a_2 + a_3 + \cdots + a_n + \cdots,$$

is called an infinite series. The nth term of an infinite series is often called a_n, and the nth partial sum is often written S_n.

Hence, in our case,

$$S_1 = a_1,$$

$$S_2 = a_1 + a_2,$$

$$\cdots,$$

$$\cdots,$$

$$\cdots,$$

$$S_n = a_1 + a_2 + a_3 + \cdots + a_n.$$

We can get greater insight into this infinite series by considering the partial sums $S_1, S_2, S_3, \ldots$ given below. Note that S_3, for example, is the sum of the first three terms.

$S_1 = (1/2)$ $\qquad = 2 - (3/2) = 2 - ((1+2)/2^1)$
$S_2 = (1/2) + (2/4) = (4/4)$ $\qquad = 2 - (4/4) = 2 - ((2+2)/2^2)$
$S_3 = (1/2) + (2/4) + (3/8) = (11/8)$ $\quad = 2 - (5/8) = 2 - ((3+2)/2^3)$
$S_4 = (1/2) + (2/4) + (3/8) + (4/16)$ $\quad = (26/16)$ $\quad = 2 - (6/16) = 2 - ((4+2)/2^4)$

What would be the average number of children per woman if all couples decided not to have more than four children, even if they still did not have a boy? The answer is S_4, the fourth partial sum, or $2 - \frac{6}{16}$ or $26/16$ children or an average of 1 and 5/8 children.

We wish to know if the partial sums approach a limit as we keep adding terms to our series (adding more children to families). To do this, we seek a general term S_n, which would give us the partial sum after n terms. Note that each partial sum involves subtracting a fraction from 2. Also, notice that the numerators vary in a linear manner and the denominators increase by a factor of 2. Can you figure out what the general term S_n is?

The result is $S_n = 2 - (n + 2)/2^n$. Check it against the partial sums for $n = 1, 2, 3$, and 4.

The sum of the infinite series is defined as the limit of S_n as n approaches infinity, if this limit exists. Can you determine what S_n approaches, as n approaches infinity? Since 2^n grows much faster than $(n + 2)$, the fraction approaches zero (see Exercise 3). Hence, S_n converges to 2, and the fertility rate is 2. (Recall that fertility rate is the average number of children per woman.) Of course, this assumes that arbitrarily large family sizes are possible; and since they aren't, the fertility rate in this sample country would be slightly less than 2, thus below the replacement-level fertility.

Example 2   Now let's consider a country in which each couple wishes to have both a son and a daughter. Each woman will have children until she has at least one boy and one girl, and then she will stop. The tree diagram for this case is shown in Figure 3.

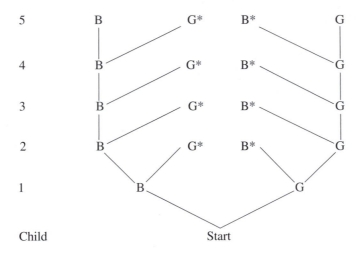

Figure 3. Tree diagram for Example 2.

Our reasoning is similar to the previous case. Notice that no women will have just one child, one-half of the women will have two children, one-quarter will have three children, one-eighth will have four children, and so on. Hence, the average number of children per woman is

$$\text{Average} = 2(1/2) + 3(1/4) + 4(1/8) + 5(1/16) + \cdots + (n + 1)/(2^n) + \cdots$$

Again, we consider the partial sums

$$S_1 = (2/2) \qquad\qquad\qquad\qquad = 3 - (4/2)$$
$$S_2 = (2/2) + (3/4) = (7/4) \qquad\qquad = 3 - (5/4)$$
$$S_3 = (2/2) + (3/4) + (4/8) = (18/8) \qquad = 3 - (6/8)$$
$$S_4 = (2/2) + (3/4) + (4/8) + (5/16) = (41/16) \quad = 3 - (7/16)$$

If all couples decided to stop after four children, the average number of children per woman would be S_4 or $3 - \frac{7}{16}$ or $2\frac{9}{16}$ children per woman.

To get the sum of our infinite series, we again obtain the general term S_n. Can you obtain S_n using reasoning similar to that used in the first example? The result is $S_n = 3 - (n+3)/2^n$. Check this against the partial sums above. Reasoning as before, we find that S_n converges to 3, and the fertility rate in this sample country will be above the replacement-level fertility.

Example 3   For a final example, we return to a country where boys are valued more highly than girls, but where infant mortality is high and many children are physically or mentally retarded due to malnutrition. To increase prospects for economic security, each woman has children until she has at least two boys and then stops. Figure 4 shows the diagram.

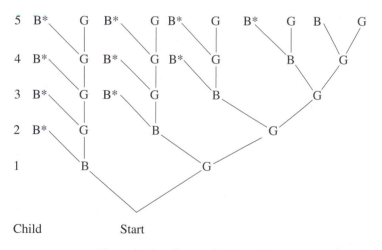

Figure 4. Tree diagram for Example 3.

Check every step in the following analysis, which uses the same approach as the previous examples. The infinite series for the average number of children per woman is

$$\text{Average} = (1/4)(2) + (2/8)(3) + (3/16)(4) + (4/32)(5) + \cdots + n(n+1)/(2^{n+1}) + \cdots$$

with partial sums

$$S_1 = (2/4)$$
$$= 4 - (14/4) = 4 - (7/2)$$
$$S_2 = (2/4) + (6/8) = (10/8)$$
$$= 4 - (22/8) = 4 - (11/4)$$
$$S_3 = (2/4) + (6/8) + (12/16) = (32/16)$$
$$= 4 - (32/16) = 4 - (16/8)$$
$$S_4 = (2/4) + (6/8) + (12/16) + (20/32) = (84/32)$$
$$= 4 - (44/32) = 4 - (22/16)$$
$$S_5 = (2/4) + (6/8) + (12/16) + (20/32) + (30/64) = (198/64)$$
$$= 4 - (58/64) = 4 - (29/32).$$

After some trial-and-error (or by using more systematic methods such as considering the second differences of the numerators), we obtain

$$S_n = 4 - \{(n+2)(n+3) + 2\}/2^{n+1},$$

which approaches 4 as n gets large. In this case, the fertility rate will be far above the replacement-level fertility. Notice that the desire to have two sons, rather than a son and a daughter, raises the fertility rate from three to four, with serious consequences for the rate of population growth.

EXERCISES (TREE DIAGRAM METHOD)

1. Draw a tree diagram to show possible families in terms of the order in which boys and girls are born when there are four children. Use the tree diagram to find the probabilities that there are

    (a) no girls;

    (b) one girl;

    (c) two girls;

    (d) three girls;

    (e) four girls.

2. Write an equation for S_n, the nth partial sum when

    (a) $S_1 = 2 - 2/4$; $S_2 = 2 - 3/8$; $S_3 = 2 - 4/16$; $S_4 = 2 - 5/32$; . . . .

    (b) $S_1 = 3 - 5/2$; $S_2 = 3 - 6/4$; $S_3 = 3 - 7/8$; $S_4 = 3 - 8/16$; . . . .

    What does S_n approach in each case when n approaches infinity?

3. Using the equation $S_n = 2 - ((n + 1)/(2^{n+1}))$, check the convergence of this series by finding S_n when n successively equals 5, 10, 20, etc., using a calculator. What conclusion can you make?

4. Consider a society where couples have children until they either have two girls or one boy. Calculate the average number of children per couple under these conditions.

5. Consider a society where couples have children until they have either two boys or three girls. Calculate the average number of children per couple under these conditions.

6. Consider a society where couples have children until they have either three boys or a total of five children. What would be the average number of children per couple in this case?

7. As indicated previously, world population in 1997 was 5.84 billion people, and was doubling every 47 years. If that doubling time remained constant, what would world population be in 235 years after 1997? What conclusions can you draw from this result?

8. The average number of children obtained for different assumptions can be simulated by tossing coins. For example, the situation in Example 1 can be simulated by tossing a coin repeatedly until a head is obtained. If a head is first obtained on a fourth toss, this is equivalent to a couple having four children: three girls and a boy. Simulate results for 30 couples; compute the average number of tosses per couple (children per couple) and compare your results with the theoretical value of 2.

9. Set up a coin simulation for the case when couples have children until they have one boy and one girl (Example 2). Again compare your results with the theoretical result of 3.

2   TABULAR METHOD

This method presents a simplified model that will help students gain insight into population growth issues. Some necessary definitions are

1. Doubling time—the number of years for a population to double.
2. Birth cycle—the number of years from a woman's birth to the beginning of her child-bearing years.

Assumptions Related to the Model

The basic assumptions for the model of population growth are:

1. A birth cycle is 20 years.
2. Each person lives 80 years.
3. The sex ratio remains constant at one female to one male.
4. The average fertility rate per woman remain fixed.

TABLE 1   Tabular Method

Age	0	20	40	60	80
Birth cycle	Childhood C	Reproductive R	Post Reproductive PR_1	Post Reproductive PR_2	Total

Table 1 shows how the lifespan is assumed to be divided into four periods of 20 years each.

C stands for the childhood period of 0 to 20 years of age. R represents the reproductive period, assumed to last from ages 20 to 40. (Assumptions are made to simplify. Demographers consider the female reproductive years to be from 15 to 49.) PR denotes the post-reproductive period, ages 40 to 80.

To keep the various periods all equal to 20 years, the post-reproductive period is divided into PR_1, ages 40 to 60, in which people become grandparents, and PR_2, ages 60 to 80, in which people become great-grandparents. These assumptions are made to simplify a very complicated problem in order to gain valuable insights into concepts related to population growth.

Use of Model for the Case When Women Have an Average of Four Children

We will first consider population growth when the average number of children per woman is four (see Table 2).

TABLE 2   Tabular Method

Age	0	20	40	60	80
Birth cycle	Childhood C	Reproductive R(Female)	Post Reproductive PR_1	Post Reproductive PR_2	Total
0	20	0	0	0	20
1	40	20(10)	0	0	60

We assume for this model that 20 children (10 males and 10 females), tired of the problems of modern society, have fled to a remote island to make a new beginning. Hence, there are no people in the other age categories. The choice of 20 is purely arbitrary; however, it is important that an even number be selected in order to have a whole number of females during the reproductive periods, as will be seen below.

Between birth cycle 0 and birth cycle 1 (in Table 2), 20 years have elapsed. The 20 original children are all 20 years older and have moved into the reproductive period, *R*. Based on the given assumptions, there are 10 males and 10 females, and the females are able to reproduce. The 10 in parenthesis in the *R* column denotes 10 females. Because the average female has four children, there are 40 children born during the 20 years of birth cycle 1. As in the previous model, the average number of children per woman and average number of children per couple are assumed to be equal.

TABLE 3 Tabular Method

Age	0	20	40	60	80
Birth cycle	Childhood C	Reproductive R(female)	Post Reproductive PR_1	Post Reproductive PR_2	Total
0	20	0	0	0	20
1	40	20(10)	0	0	60
2	80	40(20)	20	0	140
3	160	80(40)	40	20	300
4	320	160(80)	80	40	600
5	640	320(160)	160	80	1200
6	1280	640(320)	320	160	2400

Summary

1. For every birth cycle, the children of the previous birth cycle move into the reproductive period, and have an average of four children per female.

2. Those people in post-productive period 2 (PR_2) for a given cycle all die at the end of that cycle, since they reach age 80.

3. The total population of 2400 at the end of birth cycle 6 can be found in three different ways.

   (a) The total population equals the sum of the number of people in the childhood, reproductive, and post-reproductive periods.

   $$\text{Total Population} = C + R + PR_1 + PR_2$$
   $$= 1280 + 640 + 320 + 160$$
   $$= 2400 \text{ people.}$$

   (b) The population at the end of birth cycle 6 equals that at the end of birth cycle 5 plus births during birth cycle 6 minus deaths during birth cycle 6. Hence,

   $$\text{Total Population} = 1200 + 1280 - 80$$
   $$= 2400 \text{ people.}$$

   Note that population increases depend not only on births, but on deaths as well. Generally, birth rates and death rates are high in developing nations and relatively low in developed nations.

(c) The number of people in each age category doubled in going from birth cycle 5 to birth cycle 6. Thus, the total population must also double, and

$$\text{Total Population} = 2(1200)$$
$$= 2400 \text{ people.}$$

4. The percentage of people under age 20 after birth cycle 6 is

$$(1280/2400) \times 100 = 53.3\%.$$

This percentage is very high, indicative of a rapidly growing population. At the present time, in the developing countries, almost 40% of the people are under 15 years of age. When such a large percentage of a country's population are dependent, it makes it more difficult for a country to develop its economy.

5. The doubling time for population is one birth cycle, or 20 years. If no changes were made, the population would continue to double every 20 years with greater and greater impact on population, available resources, crowding, etc.

Analysis of Strategies for Reducing Population Growth

Our model has shown that the community that started with only 20 children has very rapidly grown to 2400 people. Assume that this growth is starting to strain the environment, that crowding is occurring, that the air and water are becoming polluted, and that the general quality of life is deteriorating. Further, our analysis indicates another doubling of population in 20 years, unless changes occur. Some possible changes to slow population growth and help avoid or reduce continued deterioration of the quality of life in this community are considered next.

(It should be pointed out that in addition to population, such factors as lifestyle, waste, and social and economic conditions greatly affect the environment and the quality of life.)

3 ZERO POPULATION GROWTH

Suppose someone suggests a stabilization of population, or zero population growth (ZPG), during birth cycle 7, as in Table 4.

TABLE 4 Tabular Method

Age	0	20	40	60	80
Birth cycle	Childhood C	Reproductive R(female)	Post Reproductive PR_1	Post Reproductive PR_2	Total
6	1280	640(320)	320	160	2400
7a	160	1280(640)	640	320	2400

The population at the end of birth cycle 7 (denoted here as 7a, since it is only one possibility) must remain at 2400. Hence, since 160 people die during the birth cycle, only 160 children can be born.

Related Problems

Note the many problems related to the abrupt shift from four children per woman to zero population growth:

1. The average number of children per female during this cycle is 160/640 or just one in four. Or, for every four women, on the average there will be three women who have no children at all and the other woman will have only one child. Imagine the difficulties involved in trying to obtain such a condition. However, these difficulties must be considered along with the very negative effects of not taking action, such as resource scarcities, heavy pollution, starvation, etc., showing the importance of anticipating difficulties and taking proper action before the situation becomes critical.

2. There is a tremendous decrease in the number of children from birth cycle 6 to birth cycle 7a, from 1280 children to 160, or from 53% of the population to only about 7%. Consider this in terms of the economy; teachers, manufacturers of toys, producers of baby foods, and others dependent on children for their livelihoods. Many workers depend on a high population growth rate for jobs. Such a rapid change in the number of children in the society would have a very disruptive effect on many lives.

4   SHIFT TO AVERAGE OF TWO CHILDREN PER WOMAN

Next we consider a more gradual transition to a stable population. Switch to an average of two children per couple for the next birth cycles, starting again with the results from birth cycle 6. (See Table 5.)

Note how the population continues to increase for three more birth cycles, or 60 years. It finally levels off at 5120 people, more than twice the population (2400) when the switch

TABLE 5   Tabular Method

Age	0	20	40	60	80
Birth cycle	Childhood C	Reproductive R(female)	Post Reproductive PR_1	Post Reproductive PR_2	Total
6	1280	640(320)	320	160	2400
7b	1280	1280(640)	640	320	3520
8b	1280	1280(640)	1280	640	4480
9b	1280	1280(640)	1280	1280	5120
10b	1280	1280(640)	1280	1280	5120

to two children per female occurred. This delay is largely due to the high percent (53%) of the population who were children at that time. When ZPG is reached, 25% of the population is in each of the four age categories.

The results above indicate why populations continue to rise in some countries even though the birth rate is below the replacement-level fertility of 2.1 children per female. When a country has many young people, population stabilization cannot occur rapidly without severe adverse effects.

Example 1   Two children per woman (replacement-level fertility) Let's consider another community where women have an average of two children, starting again with 20 children (10 female and 10 male) and the same basic assumptions. (See Table 6.)

TABLE 6   *Tabular Method*

Age	0	20	40	60	80
Birth cycle	Childhood C	Reproductive R(female)	Post Reproductive PR$_1$	Post Reproductive PR$_2$	Total
0	20	0	0	0	20
1	2	20(10)	0	0	40
2	20	20(10)	20	0	60
3	20	20(10)	20	20	80
4	20	20(10)	20	20	80

Zero population growth is reached by the end of birth cycle 3, in which each of the age categories has 20 people. Note the tremendous difference with the previous case and consider how it would effect the environment, economy, and quality of life. Both cases show that with an average of two children per woman, zero population growth will eventually be reached. Hence, the theoretical replacement-level fertility is two children per woman. As indicated before, in the United States and other developed countries, the actual replacement-level fertility is 2.1 to take into account those who do not reach their reproductive period. In developing countries, this value may be as high as 2.5.

Example 2   Average of 1.5 children per woman: negative population growth
Since some countries, including Russia and Germany, have stabilized their populations and gone into negative population growth, Table 7, in which the average couple has 1.5 children, should be instructive.

In this case we are starting with 1280 children and no people in any of the other age categories. Note that the population initially rises due to a very young population, but after birth cycle 3, it starts to decline and approach a population of zero. Many people in countries or ethnic groups where the birth rate is below the replacement level, have expressed concern about the effects of a negative population growth.

TABLE 7   Tabular Method

Age	0	20	40	60	80
Birth cycle	Childhood C	Reproductive R(female)	Post Reproductive PR$_1$	Post Reproductive PR$_2$	Total
0	1280	0	0	0	1280
1	960	1280(640)	0	0	2240
2	720	960(480)	1280	0	2960
3	540	720(360)	960	1280	3500
4	405	540(270)	720	960	2625
5	303	405(202)	540	720	1968

In conclusion, with the simplified model and only very basic calculations, students can explore many important concepts related to percent of population under 20 years of age, replacement-level fertility, and zero population growth. These problems can serve as the basis for more involved analyses of population growth.

EXERCISES (TABULAR METHOD)

For the problems below, use the same assumptions as in the sample examples, unless otherwise indicated.

1. Starting with 512 children and assuming that couples have an average of 2.5 children, set up a table to show how the population increases and calculate the approximate doubling time. (If the number of people in a reproductive period is odd, assume that one person does not get married.)

2. Assume that after five birth cycles, a population consists of 3200 children, 1600 people in the reproductive period, 800 people in the post-reproductive period 1, and 400 people in the post-reproductive period 2.

   Indicate how many people there would be in each age category and the total population for the next birth cycles for the following conditions for birth cycles 6, 7, and 8, respectively.

   (a) Four children per female (for birth cycle 6).
   (b) Zero population growth (for birth cycle 7).
   (c) Two children per female (for birth cycle 8).

   Note that each part is to be worked out using conditions at the end of the previous birth cycle. In addition,

   (d) What is the doubling time for birth cycle 6?
   (e) What percent of children are under 20 years of age at the end of birth cycle 7?
   (f) What is the average number of children per couple for birth cycle 7?

5   MATRIX FORM MODEL

This method presents an approach that helps us

- Gain deeper insight into population growth issues,
- Make a mathematical model based on the given conditions,
- Program the model for use of a computer or a graphing calculator.

Assumptions Related to the Model

The basic assumptions for the model of population growth are the same as those for the previous model:

1. The birth cycle is 20 years.
2. Each person lives 80 years.
3. The sex ratio remains constant at one female to one male.
4. The average fertility rate per woman remain fixed.

Definitions and Notations

Definition. *A rectangular array of numbers of the form*

$$\begin{pmatrix} a_{11} & a_{12} & \cdots & a_{1n} \\ a_{21} & a_{22} & \cdots & a_{2n} \\ \cdots & \cdots & \cdots & \cdots \\ \cdots & \cdots & \cdots & \cdots \\ a_{m1} & a_{m2} & \cdots & a_{mn} \end{pmatrix}$$

is called an $m \times n$ matrix. Here m is the number of rows and n is the number of columns of A. The element a_{ij} $(1 \leq i \leq m$ and $1 \leq j \leq n)$ *represents an element of A located at the intersection of the ith row and the jth column of A.*

Matrix Multiplication

Suppose A is an $m \times n$ matrix and B is an $n \times r$ matrix. Then the product of A and B denoted by AB is an $m \times r$ matrix C such that the element c_{ij} is obtained by the sum of products of corresponding elements of the ith row of A and the jth column of B. For example the element c_{13} can be obtained by the sum of products of corresponding elements of the 1st row of A and the 3rd column of B.

$$c_{13} = \begin{bmatrix} a_{11} & a_{12} & \cdots & a_{1n} \end{bmatrix} \begin{bmatrix} b_{13} \\ b_{23} \\ \cdots \\ b_{n3} \end{bmatrix}$$

or,

$$c_{13} = a_{11}b_{13} + a_{12}b_{23} + a_{13}b_{33} + \cdots + a_{1n}b_{n3}.$$

Population Growth Problems

Let us reconsider the population problems stated in the Tabular Method in a more general way. Assume a 1×4 matrix $A = [c \quad p \quad pr_1 \quad pr_2]$, where c stands for the number of females in the childhood period of 0 to 20 years of age (group 1), p represents the number of females in the reproductive period (group 2), from ages 20 to 40. Assumptions are made to simplify. pr_1 and pr_2 denote the number of females in the post-reproductive periods (groups 3 & 4), ages 40 to 60 and 60 to 80, respectively. Each person lives 80 years.

Consider a 4×4 matrix B as follows:

$$B = \begin{bmatrix} r_1 & r_2 & 0 & 0 \\ 0 & 0 & r_3 & 0 \\ 0 & 0 & 0 & r_4 \\ 0 & 0 & 0 & 0 \end{bmatrix}$$

Here r_1 in column one represents the rate by which the number of females in group 1 of the next cycle will increase due to birth and/or immigration, where zeros show that there will be no children from other age groups. Also, r_2 in column two, r_3 in column three, and r_4 in column four show the rate by which the future number of females in age groups 2, 3, and 4, respectively, will increase due to a transfer from lower age groups and/or immigration.

$$B = \begin{bmatrix} r_1 & r_2 & 0 & 0 \\ 0 & 0 & r_3 & 0 \\ 0 & 0 & 0 & r_4 \\ 0 & 0 & 0 & 0 \end{bmatrix}$$

Note 1. If no immigration is considered then r_1 is half the number of children (half boys and half girls) per female in the reproductive stage, where r_2, r_3, and r_4 are all equal to one. Thus, in this case

$$B = \begin{bmatrix} r_1 & 1 & 0 & 0 \\ 0 & 0 & 1 & 0 \\ 0 & 0 & 0 & 1 \\ 0 & 0 & 0 & 0 \end{bmatrix}$$

Calculation

Now, assuming there are initially no people in groups 2, 3, and 4 as in the Tabular Model, we have

$$p = pr_1 = pr_2 = 0$$

and

$$A = [c \quad 0 \quad 0 \quad 0]$$

where c is the number of females initially present. Then the number of females in the different age groups in the 1st, 2nd, 3rd, ..., and nth cycles can be obtained simply by $(A)B = AB$, $(AB)B = AB^2$, $(AB^2)B = AB^3$, ..., and AB^n respectively.

Example 1 ***Four children per female with no immigration*** Assuming no immigration, find the population growth when the average number of children per woman is four. Since there are four children per female, there are two daughters. Thus,

$$r_1 = 4/2 = 2, \quad r_2 = 1, \quad r_3 = 1, \quad r_4 = 1.$$

For comparison with the tabular model, in this model one represents ten people, two represents 20 people, etc. Hence, by starting with ten females, to simplify the analysis and assuming the $c = 1$ in matrix A represents 10 we have,

$$A = [1 \quad 0 \quad 0 \quad 0]$$

and

$$B = \begin{bmatrix} 2 & 1 & 0 & 0 \\ 0 & 0 & 1 & 0 \\ 0 & 0 & 0 & 1 \\ 0 & 0 & 0 & 0 \end{bmatrix}$$

The first cycle female-population matrix G_1 (four age categories), is given by:

$$G_1 = AB = [1 \quad 0 \quad 0 \quad 0] \begin{bmatrix} 2 & 1 & 0 & 0 \\ 0 & 0 & 1 & 0 \\ 0 & 0 & 0 & 1 \\ 0 & 0 & 0 & 0 \end{bmatrix}$$

$$= [2 \quad 1 \quad 0 \quad 0]$$

Note. Since we assumed the number of males and females are the same, to find the total number of people of the first cycle T_1, simply multiply the sum of numbers of people of all age groups in the cycle by two as follows:

$$T_1 = 2(2 + 1 + 0 + 0) = 6 \quad \text{or} \quad 60 \text{ people}$$

The number of females in the second cycle G_2 can be obtained either by

$$G_2 = G_1 B = [2 \quad 1 \quad 0 \quad 0] \begin{bmatrix} 2 & 1 & 0 & 0 \\ 0 & 0 & 1 & 0 \\ 0 & 0 & 0 & 1 \\ 0 & 0 & 0 & 0 \end{bmatrix}$$

$$= [4 \quad 2 \quad 1 \quad 0],$$

or, by

$$G_2 = AB^2 = [1 \quad 0 \quad 0 \quad 0] \begin{bmatrix} 4 & 2 & 1 & 0 \\ 0 & 0 & 0 & 1 \\ 0 & 0 & 0 & 0 \\ 0 & 0 & 0 & 0 \end{bmatrix}$$

$$= [4 \quad 2 \quad 1 \quad 0]$$

This means that at the end of birth cycle 2, there are 4 females in the childhood period, 2 females in the reproductive period, and 1 female in post-reproductive 1 period.

The total number of people in this cycle is

$$T_2 = 2(4 + 2 + 1 + 0) = 14 \quad \text{or} \quad 140 \text{ people}$$

Also, in the same way (four age categories)

$$G_3 = AB^3 = [8 \quad 4 \quad 2 \quad 1] \quad = [2^3 \quad 2^2 \quad 2^1 \quad 2^0]$$
$$G_6 = AB^6 = [64 \quad 32 \quad 16 \quad 8] = [2^6 \quad 2^5 \quad 2^4 \quad 2^3]$$

and the number of females in the 20th cycle G_{20} (four age categories), is

$$G_{20} = AB^{20} = [2^{20} \quad 2^{19} \quad 2^{18} \quad 2^{17}] = 2^{17}[8 \quad 4 \quad 2 \quad 1]$$

The total number of people in this cycle is

$$T_{20} = 2^{18}(8 + 4 + 2 + 1) = 3{,}932{,}160 \quad \text{or} \quad 39{,}321{,}600 \text{ people}$$

Thus, if we start with 10 females, then the population after 400 years will reach to 39,321,600 people. It is instructive to compare this result with those obtained with the Tabular Model (see Table 3).

Generalization of the Case

Let us solve the problem in a more general way. Consider two matrices A and B as follows:

$$A = [c \quad 0 \quad 0 \quad 0]$$

$$B = \begin{bmatrix} r & 1 & 0 & 0 \\ 0 & 0 & 1 & 0 \\ 0 & 0 & 0 & 1 \\ 0 & 0 & 0 & 0 \end{bmatrix}$$

The first cycle female-population matrix groups, G_1 is

$$G_1 = AB = [cr \quad c \quad 0 \quad 0]$$

The powers of the matrix B are:

$$B^2 = \begin{bmatrix} r & 1 & 0 & 0 \\ 0 & 0 & 1 & 0 \\ 0 & 0 & 0 & 1 \\ 0 & 0 & 0 & 0 \end{bmatrix} \begin{bmatrix} r & 1 & 0 & 0 \\ 0 & 0 & 1 & 0 \\ 0 & 0 & 0 & 1 \\ 0 & 0 & 0 & 0 \end{bmatrix} = \begin{bmatrix} r^2 & r & 1 & 0 \\ 0 & 0 & 0 & 1 \\ 0 & 0 & 0 & 0 \\ 0 & 0 & 0 & 0 \end{bmatrix}$$

$$B^5 = \begin{bmatrix} r^5 & r^4 & r^3 & r^2 \\ 0 & 0 & 0 & 0 \\ 0 & 0 & 0 & 0 \\ 0 & 0 & 0 & 0 \end{bmatrix} \quad \text{(Check it)}$$

and

$$B^n = \begin{bmatrix} r^n & r^{n-1} & r^{n-2} & r^{n-3} \\ 0 & 0 & 0 & 0 \\ 0 & 0 & 0 & 0 \\ 0 & 0 & 0 & 0 \end{bmatrix} \quad \text{for } n \geq 3,$$

and

$$G_n = AB^n = [cr^n \quad cr^{n-1} \quad cr^{n-2} \quad cr^{n-3}]$$

or

$$= c[r^n \quad r^{n-1} \quad r^{n-2} \quad r^{n-3}]$$

(Check it for $n = 3, 4, 5$.)

Total Population (A General Formula)

Since we assumed the number of males and females are the same, to find the total number of people of ith cycle T_i simply multiply the sum of numbers of people of all age groups in the ith cycle by two as follows:

$$T_0 = 2(c + 0 + 0 + 0) \qquad = 2c,$$
$$T_1 = 2(cr + c + 0 + 0) \qquad = 2c(1 + r),$$
$$T_2 = 2(cr^2 + cr + c + 0) \quad = 2c(1 + r + r^2),$$
$$T_3 = 2(cr^3 + cr^2 + cr + c) = 2c(1 + r + r^2 + r^3),$$

and

$$T_5 = 2(cr^5 + cr^4 + cr^3 + cr^2) = 2c(r^2 + r^3 + r^4 + r^5) = 2cr^2(1 + r + r^2 + r^3).$$

Hence, by the above pattern, the general term T_n will be

$$T_n = 2(cr^n + cr^{n-1} + cr^{n-2} + cr^{n-3})$$
$$= 2cr^{n-3}(1 + r + r^2 + r^3) \quad \text{for } n \geq 3,$$

or,

$$T_n = 2cr^{n-3}(1+r)(1+r^2) \quad \text{for } n \geq 3.$$

(Check it for $c = 10, 20$, and $r = 1, 2$.)

Note 2.   For $c = 1, r = 2$, and $n = 5, 6$, and 10 we have:

$$T_5 = 2(1)2^{5-3}(1+2)(1+2^2) = 2(1)(4)(3)(5) = 120,$$
$$T_6 = 2(1)2^{6-3}(1+2)(1+2^2) = 2(1)(8)(3)(5) = 240,$$
$$T_{10} = 2(1)2^{10-3}(1+2)(1+2^2) = 2(1)(128)(3)(5) = 3840.$$

Note 3.   We observe that for fixed c and r, the total population after the second cycle will follow an exponential path.

$$T_n = [2c(1+r)(1+r^2)]r^{n-3} = kr^{n-3} \quad \text{for } n \geq 3,$$

where

$$k = 2c(1+r)(1+r^2) \quad \text{is a constant.}$$

Note 4.   The ratio of T_n to T_{n-1} after the second cycle is given by

$$T_n/T_{n-1} = [2c(1+r)(1+r^2)]r^{n-3}/[2c(1+r)(1+r^2)]r^{(n-1)-3}$$
$$= r \quad \text{for } n \geq 3.$$

This means the total population after the second cycle is r times the population of the previous cycle.

Note 5.   Since

$$B^n = \begin{bmatrix} r^n & r^{n-1} & r^{n-2} & r^{n-3} \\ 0 & 0 & 0 & 0 \\ 0 & 0 & 0 & 0 \\ 0 & 0 & 0 & 0 \end{bmatrix} \quad \text{for } n \geq 3,$$

then for any matrix

$$A = [c \quad d \quad e \quad f], \quad AB^n = [cr^n \quad cr^{n-1} \quad cr^{n-2} \quad cr^{n-3}]$$

or

$$= c[r^n \quad r^{n-1} \quad r^{n-2} \quad r^{n-3}]$$

Again, compare these results with those obtained by the Tabular Method.

Example 2   Switch from four to average of two children per female

In Example 1, suppose after 10 cycles the people switch from four children per female to an average of two children per female for the next birth cycles, to stabilize the population. Show the results for the next 5 birth cycles.

First we need to find the 10th birth cycle matrix G_{10} (four age categories). To do so, let $r = 4/2 = 2$, $c = 1$, $n = 10$, and use the general formula on page 136—

$$G_n = c[r^n \quad r^{n-1} \quad r^{n-2} \quad r^{n-3}].$$

Thus, the tenth cycle female-population matrix groups (four age categories), G_{10} is

$$G_{10} = 1[2^{10} \quad 2^{10-1} \quad 2^{10-2} \quad 2^{10-3}] = [1024 \quad 512 \quad 256 \quad 128]$$

and the total population in 10th cycle T_{10} is

$$T_{10} = 2(1024 + 512 + 256 + 128) = 3840 \quad \text{or } 38400 \text{ people}$$

Since we are switching to an average of two children per female, $r = r_1 = 2/2 = 1$, thus,

$$A = [1024 \quad 512 \quad 256 \quad 128],$$

$$B = \begin{bmatrix} 1 & 1 & 0 & 0 \\ 0 & 0 & 1 & 0 \\ 0 & 0 & 0 & 1 \\ 0 & 0 & 0 & 0 \end{bmatrix}$$

(Hint: enter the matrices A and B on your calculator. Multiply A by B and the results by B and continue until you get all results respectively.) Thus, the 11th cycle female-population matrix groups G_{11} (four age categories), is

$$G_{11} = AB = [1024 \quad 512 \quad 256 \quad 128] \begin{bmatrix} 1 & 1 & 0 & 0 \\ 0 & 0 & 1 & 0 \\ 0 & 0 & 0 & 1 \\ 0 & 0 & 0 & 0 \end{bmatrix}$$

$$= [1024 \quad 1024 \quad 512 \quad 256]$$

and the total population in 11th cycle T_{11} is

$$T_{11} = 2(1024 + 1024 + 512 + 256) = 5{,}632 \quad \text{or } 56{,}320 \text{ people.}$$

The 12th cycle female-population matrix groups, G_{12} is

$$G_{12} = AB^2 = [1024 \quad 1024 \quad 1024 \quad 512]$$

and the total population in 12th cycle T_{12} is

$$T_{12} = 2(1024 + 1024 + 1024 + 512) = 7{,}168 \quad \text{or } 71{,}680 \text{ people.}$$

The 13th cycle female-population matrix groups, G_{13} is

$$G_{13} = AB^3 = [1024 \quad 1024 \quad 1024 \quad 1024]$$

and the total population in 13th cycle T_{13} is

$$T_{13} - 2(1024 + 1024 + 1024 + 1024) - 8,192 \quad \text{or } 81,920 \text{ people}$$

The 14th cycle female-population matrix groups G_{14} (four age categories), is

$$G_{14} = AB^4 = [1024 \quad 1024 \quad 1024 \quad 1024]$$

and the total population in 14th cycle T_{14} is

$$T_{14} = 2(1024 + 1024 + 1024 + 1024) = 8,192 \quad \text{or } 81,920 \text{ people.}$$

Note how the population continues to increase for three more birth cycles, or 60 years. It finally levels off at 81920 people, more than twice the population (38400) when the switch to two children per female occurred. This delay is largely due to the high percent (53%) of the population who were children at that time. When ZPG is reached, 25% of the population is in each of the four age categories. Note that this checks results found using the Tabular Method.

EXERCISES (MATRIX METHOD)

In all following problems, use the pattern and assumptions of the Matrix Method and assume no immigration.

1. The average imported quarter-pound fast food hamburger patty requires the destruction of 55 square feet of tropical forest for grazing. If each person in the world on the average eats one imported quarter-pound hamburger per day, how many square feet of tropical forest for grazing will be destroyed in the year 2096 if the world's female population matrix in 1996 was [978   782   625   500] million and on the average the fertility rate per female is

    (a) 4

    (b) 3

    (c) 2

2. A nonvegetarian diet requires about 3.5 acres/person. Assuming the fertility rate in the U.S. is 2.1 per female and the female population matrix in 1996 was [36   34   32   31] million, how many acres must be devoted to feed the population of the U.S. in the

    (a) fourth cycle?

    (b) sixth cycle?

3. Using similar data as given in Exercise 2, if a total vegetarian diet requires only about one-fifth of an acre, how many acres would be needed to feed the same population?

4. The production of one pound of steak uses 2,500 gallons of water in California. Assume each person on the average uses half a pound of steak per day and the female population matrix in 1996 was [36 34 32 31] million. How many gallons of water would be used to prepare steak in one day in California for the people in 2056, if the average fertility rate per female is:

   (a) 2.1?

   (b) 2.5?

5. In the United States the production of a pound of steak (500 calories of food energy) uses 20,000 calories of fossil fuel. Assume the U.S. female population matrix in 1996 was [36 34 32 31] million and each person uses on the average half a pound of steak per day. Calculate the amount of calories of fossil fuels that will be consumed to make steak in one day for the United States population in 2076 if the fertility rate on the average per female is:

   (a) 2.1?

   (b) 3?

Other Related Environmental Concerns

It is important that people take rapid population growth more seriously, but we should look at other aspects of the problem. The current crises are not due only to overpopulation; there are other important issues.

What the world needs today even more than ZPG is ZPIG, zero population-impact growth. For it is not just the number of people that is important, but how much they produce, consume, and waste. Affluent nations have an impact on the environment very disproportionate to their populations. The United States, with less than 5% of the world's population, uses a third of the world's resources and causes almost half of its industrial pollution. It has been estimated that an average American has 50 times the environmental impact of an average person in poor countries, in terms of resources used and pollution caused. This means that the U.S. 1997 population of 268 million people has a negative effect on ecosystems equal to over 13 billion Third World people, or well over twice the world's population.

It should be noted that wastefulness, injustice, and inequitable distribution of resources also contribute to current world problems. Most Americans connect the widespread hunger in the world today to overpopulation. Yet, several studies have indicated that there is currently enough food in the world to feed all the world's people adequately, and the problem lies in waste and inequitable distribution. For example, in the U.S. over 70 percent of the grain produced goes to feed the 9 billion animals destined for slaughter each year, and two-thirds of our grain exports are used for animal feed, while almost a billion of the world's people lack enough food. For every ton of grain produced in the U.S., how many pounds are fed to animals? How many animals are killed every minute in the U.S., on the average?

For further information on basic demographic definition, equations, and relationships and/or population-related problems, the following resources should be helpful.

REFERENCES

Ehrlich, Paul and Ann Ehrlich. 1970. *Population, Resources, Environment-Issues in Human Ecology.* San Francisco: W.H. Freeman & Co.

Miles, R., Jr. Man's Population Predicament. *Bulletin of the Population Reference Bureau.* Vol. 27, No. 2, 1337 Connecticut Avenue, NW, Washington, DC 20036.

Schaefer, L. An Introduction to Population, Environment, Society—A Teacher's Reference Manual. *Environment-Population Education Service*, 21 Merritt Street, Hamden, CT 06511. (This book provides a brief introduction to the tabular population model.)

Schwartz, R. H. *Mathematics and Global Survival.* Needham Heights, Massachusetts: Ginn Publishing, 1993 (Third Edition).

Schwartz, R. H. "Revitalizing Liberal Arts Mathematics," *Mathematics and Computer Education Journal*, Fall, 1992, 272–277.

Schwartz, R. H. "Population Growth, Tree Diagrams and Infinite Series," *UMAP Journal*, March, 1985, 35–40.

Schwartz, R. H. "A Simple Mathematical Model for Population Growth," *Journal of Environmental Education*, Vol. 12, No. 2, Winter 1980–81, 38–41.

The Population Reference Bureau, Inc. (Suite 520, 1875 Connecticut Ave. NW, Washington DC 20009) has a wide variety of material related to population. Especially valuable are the annual World Population Data Sheet and "Population Sheets," which contain graphs, charts, and discussions of issues.

Age Structured Population Models

William D. Stone
New Mexico Institute of Mining & Technology

1   INTRODUCTION

In this module we will consider several ways to model population growth in birds. (Of course, similar methods could be used for other types of animals.) We will start with a very simple linear model, then using information about birds we will build an age-structured population model. After analyzing our linear model we will consider several nonlinear effects, and how to model and analyze them. Finally, we will consider how to add some random effects.

This sort of population modeling has many uses. We can get a better understanding of the system by modeling it, and we can examine the effects of different parameters. Models are used to make wildlife management decisions. Further, modeling can help determine what information is important, and thus influence field work.

2   A VERY SIMPLE MODEL

First, let us consider a species that reproduces once a year. There is some average number of offspring per female, and some average fraction of animals that survive the year. In many species the number of males and females are approximately equal, so we will keep track only of the females.

If each female gives birth to an average of b females per year and a fraction of the population s survive the year, we can calculate the next year's population from this year's:

$$P_{n+1} = s(1 + b)P_n$$

where P_n = the female population in the nth year. If P_0 is the population we start with, after one year we have

$$P_1 = s(1+b)P_0.$$

A year later we have

$$P_2 = s(1+b)P_1 = [s(1+b)]^2 P_0.$$

Try a couple more steps. The pattern becomes clear,

$$P_n = [s(1+b)]^n P_0.$$

Example 2.1 Suppose a nest survey shows an average of 3.6 eggs per nest, and banding data indicates a 50% survival rate each year. What happens to the population size? Assuming an equal number of males and females, we have a female birth rate $b = 1.8$, with a survival rate $s = .5$. Thus

$$P_{n+1} = .5(1+1.8)P_n = 1.4P_n$$
$$P_1 = 1.4P_0, \ P_2 = 1.4P_1 = 1.96P_0,$$
$$\vdots$$
$$P_5 = (1.4)^5 P_0 = 5.376P_0.$$

Our population is growing.

Example 2.2 If a population has a female birth rate of 1.1 and a survival rate of 45%, what happens to the population size?

$$b = 1.1 \quad \text{and} \quad s = .45$$

gives

$$P_{n+1} = (.45)(1+1.1)P_n = .945P_n$$

Each population is smaller than the one before; our population is decreasing.

EXERCISES

2.1. Suppose the average number of female births is 1.5, and 75% of the population survives each year. What happens to the population size?

2.2. Suppose the average number of female births is 1.2 and 40% of the population survives each year. What happens to the population size?

2.3. What is the critical difference in problems 2.1 and 2.2?

2.4. Suppose you have a population where 60% survive each year. If you start with 100 newborn individuals, how many live to be one year old? How many to two? To three? Approximately, what is the average life span of these animals?

3 THE BIRDS AND THE BEES, FOR BIRDS

The simple model we developed in the previous section may be all right for some species, but many have a more complicated life cycle. Many bird species have three distinct life stages: chicks, juveniles, and adults.

A chick is a young bird, still being tended by her parents. The period of being a chick varies considerably among different species of birds. We will consider a species that stay with their parents for one full season.

When chicks are ready to leave their parents, they become juveniles. Although not ready to reproduce, they are no longer being tended by their parents. In some species, juveniles have the lowest survival rates; their parents are not looking after them the way the chicks' parents are, but the juveniles haven't established their territory and learned how to survive the way the adults have.

Juveniles who survive a season become adults. Adults have a relatively high survival rate and, of course, lay eggs and raise chicks.

It seems, then, that we need four numbers: the average number of chicks per adult, and the survival rates of chicks, juveniles, and adults. Biologists measure these rates by counting chicks in the nest and doing cohort studies. We will consider this in the next section.

Once we have birth rates and survival rates, we can construct a basic linear model. The number of chicks next year will be the per capita birth rate times the number of adults this year. The number of chicks this year times the survival rate for chicks gives next year's juveniles. The adults next year will be this year's adults that survive plus the juveniles that survive. Our model is

$$C_{n+1} = bA_n$$

$$J_{n+1} = s_C C_n$$

$$A_{n+1} = s_J J_n + s_A A_n.$$

Example 3.1 Given the data

	Year 1	Year 2	Year 3
Chicks	110	120	140
Juveniles	40	65	70
Adults	50	55	75

estimate b, s_C, s_J, and s_A.

The first season, 50 adults produced 120 chicks or $\frac{120}{50} = 2.4$ chicks per adult. The next season we had $\frac{140}{55} = 2.55$ chicks per adult. Averaging we estimate $b \approx 2.47$.

The first season $\frac{65}{110}$ chicks survived to become juveniles, the next season, $\frac{70}{120}$. Averaging gives $s_C \approx .59$.

Survival rates for juveniles and adults are trickier, since the next year both are adults. Going from year 1 to year 2, we have

$$40s_J + 50s_A = 55.$$

Similarly, the next year gives

$$65s_J + 55s_A = 75.$$

Solving simultaneously gives $s_J = .69$, $s_A = .55$.

EXERCISES

3.1. Suppose you are given the following population data.

	Year 1	Year 2	Year 3
Chicks	200	210	220
Juveniles	120	120	125
Adults	105	110	114

Estimate b, s_C, s_J, and s_A.

3.2. How would this model change for species that are adults immediately after their year of being a chick?

4   POPULATION MEASUREMENT

How does one measure a population of birds? Going out and counting, you don't know if you saw them all, and you don't always know if you've counted the same one more than once.

One method that field biologists use to count a population is called *mark and release counting*. A sample of the population is captured and tagged. In the case of birds, they are ringed by attaching light metal rings to the bird's foot. This sample is counted and released. Then another sample is captured. The fraction of marked individuals in the new sample should approximate the fraction of marked individuals in the entire population.

For example, suppose we catch and mark 276 individuals. Next time, we catch 315 individuals, 24 of which are marked. From our second sample $\frac{24}{315}$, or about 7.6% are marked. If 7.6% of our whole population is marked, then 276 is 7.6% of the whole population. We can estimate our population, then, by $\frac{276}{.076} = 3623$. To improve our estimate, we could mark the remaining 291 individuals we have captured and repeat the process.

By counting or estimating the number of individuals of different ages each year, we can develop a cohort study. Starting with a group of individuals we keep track of how many are alive after one year, two years, etc. From this sort of information we derive our survival rates.

Example 4.1 Suppose we start with 200 chicks. Each year 80% of the chicks survive, 60% of the juveniles, and 75% of the adults. What happens to the cohort?

Year 1		200 Chicks
Year 2	$200 \times s_C = 160$	Juveniles
Year 3	$160 \times s_J = 96$	Adults
Year 4	$96 \times s_A = 72$	Adults
Year 5	$72 \times s_A \cong 54$	Adults
Year 6	$54 \times s_A \cong 41$	Adults
Year 7	$41 \times s_A \cong 31$	Adults

$$\vdots$$

EXERCISES

4.1. Five cohorts, each starting with 200 individuals were followed for four years. From the data, estimate s_C, s_J, s_A.

	Cohort 1	Cohort 2	Cohort 3	Cohort 4	Cohort 5
Year 1	118	120	120	122	115
Year 2	35	36	35	40	31
Year 3	28	29	28	30	25
Year 4	22	23	21	23	19

4.2. This exercise requires, for each group of students, 1–2 cups of dry beans (light colored) and markers. Take your population of beans and capture some (aim for about 10%). Tag your captured beans with a dot from your marker. Count and return the tagged individuals to the pile and mix. Capture another sample and estimate the total number of beans. Tag the unmarked members of your new sample and repeat. Use another color marker and repeat the entire process. Finally, count your beans and compare your four estimates with the actual population.

5 THE LINEAR MODEL

We return to our linear model:

$$C_{n+1} = bA_n$$
$$J_{n+1} = s_C C_n$$
$$A_{n+1} = s_J J_n + s_A A_n$$

AGE STRUCTURED POPULATION MODELS

or in matrix form:

$$\begin{pmatrix} C \\ J \\ A \end{pmatrix}_{n+1} = \begin{pmatrix} 0 & 0 & b \\ s_C & 0 & 0 \\ 0 & s_J & s_A \end{pmatrix} \begin{pmatrix} C \\ J \\ A \end{pmatrix}_n \tag{1}$$

Example 5.1 Considering again the parameters we derived in example 3.1, in matrix form we have

$$\begin{pmatrix} C \\ J \\ A \end{pmatrix}_{n+1} = \begin{pmatrix} 0 & 0 & 2.47 \\ .59 & 0 & 0 \\ 0 & .69 & .55 \end{pmatrix} \begin{pmatrix} C \\ J \\ A \end{pmatrix}_n$$

Starting with 50 individuals in each age group, and multiplying by our matrix we get

$$\begin{pmatrix} C \\ J \\ A \end{pmatrix}_0 = \begin{pmatrix} 50 \\ 50 \\ 50 \end{pmatrix}, \begin{pmatrix} C \\ J \\ A \end{pmatrix}_1 = \begin{pmatrix} 124 \\ 30 \\ 62 \end{pmatrix}, \begin{pmatrix} C \\ J \\ A \end{pmatrix}_2 = \begin{pmatrix} 153 \\ 73 \\ 54 \end{pmatrix},$$

$$\dots, \begin{pmatrix} C \\ J \\ A \end{pmatrix}_{10} = \begin{pmatrix} 649 \\ 315 \\ 331 \end{pmatrix}$$

clearly, our population is growing.

Example 5.2 Considering the same model we use in example 5.1, we obtain the stable age distribution.

$$\frac{1}{P_0}\begin{pmatrix} C \\ J \\ A \end{pmatrix}_0 = \begin{pmatrix} .33 \\ .33 \\ .33 \end{pmatrix}, \frac{1}{P_1}\begin{pmatrix} C \\ J \\ A \end{pmatrix}_1 = \begin{pmatrix} .57 \\ .14 \\ .29 \end{pmatrix}, \frac{1}{P_2}\begin{pmatrix} C \\ J \\ A \end{pmatrix}_2 = \begin{pmatrix} .55 \\ .26 \\ .19 \end{pmatrix},$$

$$\dots, \frac{1}{P_{10}}\begin{pmatrix} C \\ J \\ A \end{pmatrix}_{10} = \begin{pmatrix} .50 \\ .24 \\ .26 \end{pmatrix}, \dots, \frac{1}{P_{20}}\begin{pmatrix} C \\ J \\ A \end{pmatrix}_{20} = \begin{pmatrix} .51 \\ .24 \\ .25 \end{pmatrix}$$

EXERCISES

5.1. Let $b = 2$, $s_C = .6$, $s_J = .3$, $s_A = .7$ in equation (1). Choose a reasonable starting value (no negative populations, at least some birds to start with) and run the model for several generations. What is happening to your population?

5.2. Repeat the previous exercise with $b = 1.5$, $s_C = .6$, $s_J = .2$, $s_A = .7$.

As before, we see that populations can grow or decline. Of course things are not as simple as with our basic model, but we can still consider the overall growth rate. Let

$P_n = C_n + J_n + A_n$. The ratio P_{n+1}/P_n is not a constant as it was in section 2, but it approaches one.

5.3. Consider the ratio P_{n+1}/P_n as n gets large for the models in Exercises 5.1 and 5.2. What do these ratios approach?

5.4. Compare your results in Exercise 5.3 to the results of someone who used a different starting population than you in Exercises 5.1 and 5.2.

More is going on now than just the population growth; the population distribution is changing. If we divide each vector of population by the total population,

$$\frac{1}{P_n} \begin{pmatrix} C_n \\ J_n \\ A_n \end{pmatrix},$$

we can see what happens to the age distribution.

5.5. Consider the age distributions for the models in 5.1 and 5.2. What happens as n gets large?

5.6. Compare your results in 5.5 to the results of someone who used a different starting population than you in 5.1 and 5.2.

What we have just computed is called the *stable age distribution* for the population.

6   DECREASED EGG PRODUCTION

The simple model we considered in section 2 had three basic possible behaviors: Unbounded growth, decay to zero, or (if $s(1 + b)$ was exactly one) constant. This third situation is structurally unstable. Any change in the coefficients will almost certainly move the system into one of the other two cases. Since we only have estimates on birth rates and survival rates, this means our solution is very unreliable in this situation.

Despite this, most real-world populations seem to settle down to a constant level if left alone. If we were to suddenly double the population, there wouldn't be enough food and the survival rate would go down. In other words, we expect that the survival rates s_C, s_J, and s_A, and possibly the birth rate, b, are not constants, but functions of the population size.

Our model will still give us C_{n+1}, J_{n+1}, and A_{n+1} as functions of C_n, J_n, and A_n, but the functions will now be more complicated than the simple linear functions we had before. Such a system is called *nonlinear*. Analyzing nonlinear systems is a bit more complicated than linear systems, but numerical simulation is not much different.

To improve our model, then, we must consider how birth rates and survival rates will be affected by population size. These effects vary from species to species. In some kinds of birds, the juveniles are not completely on their own; they still depend partially on their parents. In these cases, egg production is often inhibited in parents with juveniles.

We will assume, then, that the number of (female) eggs produced per (female) adult is a function of the number of juveniles. We expect this to be a decreasing function, with a maximum at $J = 0$.

Figure 1.

If we say the average clutch for a pair of adults with no juveniles is 4 eggs (2 females) then we expect b to be 2 when $J = 0$.

What else do we know about $b(J)$? If we had lots of data we could consider various functions. Such data, however, is not always easy to get. Lacking any reason or data for a fancier model, we will approximate $b(J)$ with a linear function, $b(J) = b_0 - mJ$. A few data points would be enough to approximate m.

Example 6.1   Suppose $b(J) = 2 - .02J$, $s_C = .7$, $s_J = .2$, and $s_A = .8$. Starting with a population of 10 adults, what happens?

$$C_{n+1} = (2 - .02J_n)A_n$$

$$J_{n+1} = .7C_n$$

$$A_{n+1} = .2J_n + .8A_n$$

$$\begin{pmatrix} C \\ J \\ A \end{pmatrix}_0 = \begin{pmatrix} 0 \\ 0 \\ 10 \end{pmatrix}, \begin{pmatrix} C \\ J \\ A \end{pmatrix}_1 = \begin{pmatrix} 20 \\ 0 \\ 8 \end{pmatrix}, \begin{pmatrix} C \\ J \\ A \end{pmatrix}_2 = \begin{pmatrix} 16 \\ 14 \\ 6.4 \end{pmatrix},$$

$$\begin{pmatrix} C \\ J \\ A \end{pmatrix}_3 = \begin{pmatrix} 11 \\ 11.2 \\ 7.9 \end{pmatrix}, \ldots, \begin{pmatrix} C \\ J \\ A \end{pmatrix}_{20} = \begin{pmatrix} 22.4 \\ 15.4 \\ 13.6 \end{pmatrix}, \ldots,$$

$$\begin{pmatrix} C \\ J \\ A \end{pmatrix}_{100} = \begin{pmatrix} 40.2 \\ 28.1 \\ 28.0 \end{pmatrix}$$

The population settles down to a steady level.

EXERCISES

6.1. Ten years of data on a population of birds is given in the chart. For convenience, the data has been put in order of increasing J. Plot the data on a graph. Does it look roughly linear? Sketch a reasonable line and get its equation. Better yet, if you know linear regression, find the best line to fit the data.

Juveniles	Average Birth Rate
68	1.69
77	1.54
79	1.56
82	1.45
93	1.40
104	1.44
104	1.29
106	1.19
108	1.20
108	1.12

Thus we get our system:

$$C_{n+1} = (b_0 - m J_n) A_n$$

$$J_{n+1} = s_C C_n$$

$$A_{n+1} = s_J J_n + s_A A_n$$

6.2. Take the $b(J)$ from exercise 6.1 and use $s_C = .8$, $s_J = .3$, and $s_A = .65$ in this model. Start with a reasonable starting population and iterate 10 times. What is happening? Now iterate 100 times. What do you find?

6.3. Compare results with someone who used a different starting value.

7 DECREASED OVERALL SURVIVAL

Another possible effect of increased population is lowered survival rates across the board. This might be the case if food supply were the limiting factor.

Again, we expect the survival rates to be decreasing functions of population. Lacking any reason to the contrary, we will use linear functions. We assume that the population pressure affects all age classes equally, such that the ratios of survival rates is constant. For example, if the juvenile survival rate is .3 for very small populations, and the adult survival rate is .6, we will take the adult survival rate to be always twice the juvenile survival rate. Further, we assume that an individual has roughly the same effect on the food supply, regardless of her age class. Thus the survival rates are functions of $P_n = C_n + J_n + A_n$. Putting this together we get

$$C_{n+1} = b\,A_n$$

$$J_{n+1} = \hat{s}_C \left(1 - \frac{P_n}{k}\right) C_n \tag{2}$$

$$A_{n+1} = \hat{s}_J \left(1 - \frac{P_n}{k}\right) J_n + \hat{s}_A \left(1 - \frac{P_n}{k}\right) A_n$$

where $\hat{s}_C, \hat{s}_J, \hat{s}_A$ are the (constant) survival rates for low population, P_n is the total population, and k is a constant to be determined from field data.

Example 7.1 Assuming a model of the form in equation 2, and given the following data, estimate the parameters.

n	C	J	A	P
1	37	21	17	75
2	31	16	21	68
3	46	15	18	79
4	35	22	13	70
5	26	18	23	67
6	51	15	20	86
7	35	20	22	77
8	45	17	19	81
9	39	27	22	88
10	47	22	24	93

Using $C_{n+1} = b A_n$ we get 9 estimates for b: 1.82, 2.19, 1.94, 2, 2.22, 1.75, 2.05, 2.05, 2.14. Averaging gives 2.02.

Solving these nonlinear equations is a little tricky, since P_n/k appears nonlinearly three different places. To simplify the problem, we will solve the J_{n+1} equation for S_c and k, then use this value of k in the A_{n+1} equation to solve for S_j and S_A.

The equation $J_{n+1} = \hat{S}_C(1 - (P_n/k))C_n$ can be thought of as a linear equation for $\hat{S}_C$ and $\hat{S}_C/k$. Using our data we get

$$16 = 37 S_C - 2775\frac{S_C}{k}$$

$$15 = 31 S_C - 2108\frac{S_C}{k}$$

$$22 = 46 S_C - 3634\frac{S_C}{k}$$

$$18 = 35 S_C - 2450\frac{S_C}{k}$$

$$15 = 26 S_C - 1742\frac{S_C}{k}$$

$$20 = 51 S_C - 4386\frac{S_C}{k}$$

$$17 = 35 S_C - 2695\frac{S_C}{k}$$

$$27 = 45 S_C - 3645 \tfrac{S_C}{k}$$

$$22 = 39 S_C - 3432 \tfrac{S_C}{k}$$

In matrix form we have

$$\begin{pmatrix} 37 & 2775 \\ 31 & 2108 \\ 46 & 3034 \\ 35 & 2450 \\ 26 & 1742 \\ 51 & 4386 \\ 35 & 2695 \\ 45 & 3645 \\ 39 & 3432 \end{pmatrix} \begin{pmatrix} S_C \\ -S_C \\ k \end{pmatrix} = \begin{pmatrix} 16 \\ 15 \\ 22 \\ 18 \\ 15 \\ 20 \\ 17 \\ 27 \\ 22 \end{pmatrix}$$

When we have an over-determined equation of this sort, the least squares solution can be calculated by multiplying both sides by the transpose of the matrix, then solving the resulting system. Thus if we have

$$A \vec{x} = \vec{b}$$

we get

$$A^T A \vec{x} = A^T \vec{b}$$

then the least squares solution is

$$\vec{x} = (A^T A)^{-1} A^T \vec{b}.$$

Here, we get

$$\begin{pmatrix} S_C \\ -S_C \\ k \end{pmatrix} = \begin{pmatrix} .612 \\ -.0015 \end{pmatrix}$$

Thus $S_C = .612$, $k = 408$.

Now, using this k, we get our equations for S_J and S_A.

$$21 = 17.1 S_J + 13.9 S_A$$

$$18 = 13.3 S_J + 17.5 S_A$$

$$13 = 12.1 S_J + 14.5 S_A$$

$$23 = 18.2 S_J + 10.8 S_A$$

$$20 = 15.0 S_J + 16.7 S_A$$

$$22 = 11.8S_J + 15.8S_A$$
$$19 = 16.2S_J + 17.8S_A$$
$$22 = 13.6S_J + 15.2S_A$$
$$24 = 21.2S_J + 18.8S_A$$

The least squares solution is $S_J = .918$, $S_A = .379$.

Example 7.2 Taking the model from Example 7.1, what happens to the population? Using the parameters from Example 7.1, we have

$$C_{n+1} = 2.02A_n$$

$$J_{n+1} = .612\left(1 - \frac{P_n}{408}\right)C_n$$

$$A_{n+1} = .918\left(1 - \frac{P_n}{408}\right)J_n + .379\left(1 - \frac{P_n}{408}\right)A_n$$

Starting with 100 adults, after 10 generations we have

$$\begin{pmatrix} C \\ J \\ A \end{pmatrix} = \begin{pmatrix} 44 \\ 17 \\ 26 \end{pmatrix}$$

After 20 generations,

$$\begin{pmatrix} 45 \\ 21 \\ 22 \end{pmatrix}.$$

After 100 generations we have

$$\begin{pmatrix} 44 \\ 21 \\ 22 \end{pmatrix}$$

(rounded to integers). Our population seems to have settled down to a constant level.

EXERCISES

7.1. Take $b = 2$, $\hat{s}_C = .8$, $\hat{s}_J = .3$, $\hat{s}_A = .65$ and $k = 1000$. Repeat Exercises 6.2 and 6.3 for this model.

7.2. Assuming a model of the form (2), use linear regression to solve for the constants to fit the data below.

n	C	J	A
1	95	58	40
2	77	74	44
3	86	61	60
4	127	64	67
5	133	95	60
6	113	102	67
7	137	86	73
8	149	103	67

7.3. Take the parameters you obtained in Exercise 7.2, and repeat Exercise 7.1.

8   COMPETITION FOR NESTING SITES

In some species, the limiting factor is not food so much as room. Some birds have very specific nesting site requirements, and are very territorial. For a juvenile to survive, she must find an appropriate site, in unoccupied territory. The red cockaded woodpecker (see <http://dataadmin.irm.r9.fws.gov/bio-rcw.html>) and the northern spotted owl are two endangered species for which appropriate territory is a crucial limiting factor. Since adults normally have an established territory, the population pressure falls mainly on the juveniles, and the survival rate is mainly dependent on the number of adults. If we approximate the juvenile survival rate by a linear function of the number of adults we get:

$$C_{n+1} = b\,A_n$$
$$J_{n+1} = s_C\,C_n$$
$$A_{n+1} = \hat{s}_J(1 - kA_n)J_n + s_A A_n$$

Example 8.1   Given the following data, and assuming a model of the form given in the equation above, estimate the parameters.

n	C	J	A	P
1	196	143	103	442
2	206	115	109	430
3	229	128	96	453
4	189	140	94	423
5	185	116	96	397
6	182	121	99	402
7	201	111	103	415
8	206	122	106	434

Using $C_{n+1} = bA_n$ we get seven estimates of b: 2.0, 2.10. 1.97, 1.97, 1.90, 2.03, 2. Averaging we get 1.99.

From $J_{n+1} = S_c C_n$ we again get seven estimates of S_c. Averaging gives .615.

Setting up the equation $A_{n+1} = S_J(1 - kA_n)J_n + S_A A_n = S_j J_n - kS_j A_n J_n + S_A A_n$ for $S_J, -kS_J, S_A$ gives

$$109 = S_J \cdot 143 - kS_J \cdot 14729 + S_A \cdot 103$$

$$96 = S_J \cdot 115 - kS_J \cdot 12535 + S_A \cdot 109$$

$$94 = S_J \cdot 128 - kS_J \cdot 12288 + S_A \cdot 96$$

$$96 = S_J \cdot 140 - kS_J \cdot 13160 + S_A \cdot 94$$

$$99 = S_J \cdot 116 - kS_J \cdot 11136 + S_A \cdot 96$$

$$103 = S_J \cdot 121 - kS_J \cdot 11979 + S_A \cdot 99$$

$$106 = S_J \cdot 111 - kS_J \cdot 11433 + S_A \cdot 103$$

or in matrix form

$$\begin{pmatrix} 143 & 14729 & 103 \\ 115 & 12535 & 109 \\ 128 & 12288 & 96 \\ 140 & 13160 & 94 \\ 116 & 11136 & 96 \\ 121 & 11979 & 99 \\ 111 & 11433 & 103 \end{pmatrix} \begin{pmatrix} S_J \\ -kS_J \\ S_A \end{pmatrix} = \begin{pmatrix} 109 \\ 96 \\ 94 \\ 96 \\ 99 \\ 103 \\ 106 \end{pmatrix}.$$

Multiplying by the transpose of the matrix gives

$$\begin{pmatrix} 110076 & 10973334 & 87260 \\ 10973334 & 1096470636 & 8732666 \\ 87260 & 8732666 & 70168 \end{pmatrix} \begin{pmatrix} S_J \\ -kS_J \\ S_A \end{pmatrix} = \begin{pmatrix} 87812 \\ 8775452 \\ 70358 \end{pmatrix}$$

so the least squares solution is

$$\begin{pmatrix} S_J \\ -kS_J \\ S_A \end{pmatrix} = \begin{pmatrix} .400 \\ -.00273 \\ .844 \end{pmatrix}$$

or

$$S_J = .400$$

$$k = .0068$$

$$S_A = .844.$$

Example 8.2 Taking the parameters from Example 8.1, what happens to the population? Using the numbers from Example 8.1, we have

$$C_{n+1} = 1.99A_n$$

$$J_{n+1} = .615C_n$$

$$A_{n+1} = .40(1 - .0068A_n)J_n + .844A_n$$

starting with 100 adults only, after 10 generations we have

$$\begin{pmatrix} C \\ J \\ A \end{pmatrix} = \begin{pmatrix} 19 \\ 116 \\ 97 \end{pmatrix}.$$

After 20, we have

$$\begin{pmatrix} 199 \\ 122 \\ 100 \end{pmatrix}.$$

After 100 generations our population seems to have settled down to

$$\begin{pmatrix} 199 \\ 123 \\ 100 \end{pmatrix}$$

(rounded to integers).

EXERCISES

8.1. Using the data in Table 8.1, approximate b, s_C, $\hat{s}_J$, s_A and k.

TABLE 8.1

n	C	J	A
1	219	161	132
2	271	181	145
3	293	220	162
4	314	246	174
5	330	251	213
6	420	250	224
7	441	332	236
8	471	359	271

8.2. Using the data from Exercise 8.1, repeat Exercises 6.2 and 6.3.

8.3. What are the units on the constant k?

8.4. What are some other ways population pressure might affect a species? One example might be too many juveniles lowering the chicks' survival rate. Can you think of others? Make up a model to fit the given example or your own example.

9   FIXED POINTS

The nonlinear systems we have been considering have tended to settle down to a constant population distribution after several generations. For instance, consider the system

$$C_{n+1} = 2A_n$$

$$J_{n+1} = .8C_n \tag{3}$$

$$A_{n+1} = .4\left(1 - \frac{A_n}{500}\right)J_n + .7A_n.$$

Let $C_0 = 0 = J_n$, $A_0 = 100$; after five generations we have

$$C_5 = 202, \quad J_5 = 147, \quad A_5 = 96$$

After ten generations

$$C_{10} = 298, \quad J_{10} = 218, \quad A_{10} = 163$$

Then

$$C_{20} = 477, \quad J_{20} = 373, \quad A_{20} = 243$$

After 40 generations or so, the system has settled down to approximately

$$C = 531, \quad J = 425, \quad A = 266.$$

A constant solution to an interative system is called a *fixed point* or a steady state. Evaluating a system at $(C_n, J_n, A_n) = $ a fixed point, gives the same values for $(C_{n+1}, J_{n+1}, A_{n+1})$.

Let us look for a fixed point to system (3). Setting $C_{n+1} = C_n = C$, and similarly for J and A, we get

$$C = 2A$$

$$J = .8C$$

$$A = .4\left(1 - \frac{A}{500}\right)J + .7A$$

Substituting $C = 2A$, $J = .8C = 1.6A$, into the third equation gives

$$A = .64\left(1 - \frac{A}{500}\right)A + .7A$$

or

$$\frac{.64}{500} A^2 - .34A = 0.$$

Thus we get $A = 0$ (which implies $C = J = 0$) or

$$A = \frac{500}{.64}(.34) = 265.625,$$

so

$$C = 2A = 531.25, \quad J = .8C = 425.$$

Thus we see there are exactly two fixed points: $(0, 0, 0)$ and $(531, 425, 266)$.

Example 9.1 We had three nonlinear systems in our previous examples. Each of these converged to a constant population. Now we look for the fixed points.

(a) Example 6.1 had the system

$$C_{n+1} = (2 - .02J_n)A_n$$
$$J_{n+1} = .7C_n$$
$$A_{n+1} = .2J_n + .8A_n$$

So we look for a solution to

$$C = (2 - .02J)A$$
$$J = .7C$$
$$A = .2J + .8A$$

The third equation implies that $A = J$, so we know $A = J = .7C$ (from the second equation). Using this in the first, we get $C = (2 - .014C).7C$ so $C = A = J = 0$ or $C = 40.8$, $A = J = 28.6$. This agrees with the result in Example 6.1.

(b) Example 7.2 had the system

$$C_{n+1} = 2.02A_n$$
$$J_{n+1} = .612\left(1 - \frac{P_n}{408}\right)C_n$$
$$A_{n+1} = .918\left(1 - \frac{P_n}{408}\right)J_n + .379\left(1 - \frac{P_n}{408}\right)A_n$$

Again we eliminate subscripts and look for the fixed point, recalling that $P_n = C_n + J_n + A_n$. Eliminating C we get

$$J = 1.23624 \left(1 - \frac{J + 3.02A}{408}\right) A$$

$$A = .918 \left(1 - \frac{J + 3.02A}{408}\right) J + .379 \left(1 - \frac{J + 3.02A}{408}\right) A$$

Multiplying the first equation by A and the second by J, and eliminating the $(1 - (P/408))$ terms, gives $1.23624A^2 - .379AJ - .918J^2 = 0$. Solving this quadratic (ignoring the negative root) gives

$$A = 1.0285J.$$

Substituting this back in, we get

$$J = 1.2715 \left(1 - .01006J\right) J$$

so $J = 0 = A = C$ or $J = 21.2$, $A = 1.03J = 21.8$, $C = 2.02A = 44.1$. This also agrees with the results from Example 7.2.

(c) Example 8.2 had the system

$$C_{n+1} = 1.99A_n$$
$$J_{n+1} = .615C_n$$
$$A_{n+1} = .40 \left(1 - .0068A_n\right) J_n + .844A_n$$

Using $C = 1.99A$, and $J = .615C = 1.22A$ we get $A = .49(1 - .0068A)A + .844A$ so $.156A = .49(1 - .0068A)A$ and $A = 0 = J = C$ or $A = 100.2$, $J = 1.22A = 122.3$, $C = 199.5$. Again, this agrees with the results in the example.

EXERCISES

9.1. Find the fixed points to the system in Exercise 6.2.

9.2. Find the fixed points to the system in Exercise 7.1.

9.3. Find the fixed points to the system in Exercise 7.3.

9.4. Find the fixed points to the system in Exercise 8.1.

9.5. How did the fixed points you found compare to your solution after 100 generations?

10   STOCHASTIC MODELS

In the modeling we have done so far, all our parameters have been averages. We have taken lots of observations and then fit a number to our data. For large populations this is reasonable. Years with bigger growth and years with smaller growth balance out. For some endangered species, however, the number of individuals is so low that using average values for parameters is not very reliable. Fluctuations around the norm may have important effects. A population where a female sometimes lays 1 egg, sometimes 2 and sometimes more, is different from a population where the females always lay 2 eggs, even if the average is the same. A model that incorporates this kind of uncertainty is called a *stochastic* model.

To build a model to consider these effects, we have to consider each individual in the population, and model the significant events in their life cycle. For example, our birds are born, some survive from chicks to juveniles, some survive to adulthood, lay eggs, etc. To simplify matters, we will assume these happen at distinct times: nesting in the spring, most death (thus survival) over the winter.

Suppose for our population we expect an average of 80% of our chicks to survive to be juveniles. We need to go through our population of chicks one by one and decide whether or not they survive. We want our model to have an average survival rate for chicks of .8, though some years will be higher and some lower. To do this we choose a random number between 0 and 1. If this number is less than 0.8, the chick survives, otherwise that individual dies. After deciding whether or not each chick survives, we have the next generation's juvenile population.

Similarly, we choose a random number for each juvenile and each adult. Comparing the random number to the average survival rate, we decide whether each individual dies, or becomes part of the next generation's adults.

When it comes to determining the number of eggs, our problem gets a bit more complicated. Just knowing the average number of eggs doesn't tell us the distribution of clutch sizes: what percent of pairs have no female chicks, one, two, three or more.

If we have lots of data, we may be able to approximate this distribution. We seldom have enough information to determine how this distribution changes with population size, so we assume the distribution doesn't change. Nest surveys are made, counting the number of eggs for each nesting pair. For instance, a survey of many nests might give us the data in Table 10.1.

TABLE 10.1

Number of eggs	0	1	2	3	4	5	6
Percentage of nests	3	2	5	12	44	29	5

If we assume equal numbers of males and females, we can calculate the probabilities of no female chicks in a nest, of one female chick, of two, etc.

As an example, suppose a nest has four eggs. Each egg is equally likely to be a female or a male, so there are 16 equally likely outcomes for the nest:

MMMM   MMMF   MMFM   MMFF
MFMM   MFMF   MFFM   MFFF
FMMM   FMMF   FMFM   FMFF
FFMM   FFMF   FFFM   FFFF

Of these 16 possibilities, one has no females, four have one female, six have two females, four have three females, and one has four females. Thus the probability of no females in a nest of four eggs is $\frac{1}{16}$; the probability of one female in a nest of four eggs is $\frac{4}{16}$; and so on.

Counting up the possible outcomes for other clutch sizes we can construct Table 10.2.

TABLE 10.2 Probabilities for Various Numbers of Females

Eggs in Nest	Number of Females						
	0	1	2	3	4	5	6
0	1	0	0	0	0	0	0
1	$\frac{1}{2}$	$\frac{1}{2}$	0	0	0	0	0
2	$\frac{1}{4}$	$\frac{2}{4}$	$\frac{1}{4}$	0	0	0	0
3	$\frac{1}{8}$	$\frac{3}{8}$	$\frac{3}{8}$	$\frac{1}{8}$	0	0	0
4	$\frac{1}{16}$	$\frac{4}{16}$	$\frac{6}{16}$	$\frac{4}{16}$	$\frac{1}{16}$	0	0
5	$\frac{1}{32}$	$\frac{5}{32}$	$\frac{10}{32}$	$\frac{10}{32}$	$\frac{5}{32}$	$\frac{1}{32}$	0
6	$\frac{1}{64}$	$\frac{6}{64}$	$\frac{15}{64}$	$\frac{20}{64}$	$\frac{15}{64}$	$\frac{6}{64}$	$\frac{1}{64}$

EXERCISES

10.1. Can you see the pattern in Table 10.2? (Hint: Think about Pascal's triangle.)

Now we are ready to calculate the distribution of numbers of females. For example, given the information in Table 10.1 and Table 10.2, what is the probability of exactly 2 female chicks in a nest?

We take the probability of 2 females in a nest of k eggs times the probability of a nest with k eggs, and add them up:

$$0 \cdot \frac{3}{100} + 0 \cdot \frac{2}{100} + \frac{1}{4} \cdot \frac{5}{100} + \frac{3}{8} \cdot \frac{12}{100} + \frac{6}{16} \cdot \frac{44}{100} + \frac{10}{32} \cdot \frac{29}{100} + \frac{20}{64} \cdot \frac{5}{100} = .3248$$

In the same way, we calculate the probabilities of other numbers of females in a nest, and obtain Table 10.3.

TABLE 10.3

No. of female chicks	0	1	2	3	4	5	6
Probability	.105	.24	.325	.231	.084	.014	.001

10.2. Calculate Table 10.3 from Table 10.1 and Table 10.2.

Finally we simulate the number of chicks in our next generation. For each adult female choose a random number. If it is less than .105, this bird has no female chicks this year. If the random number is between .105 and .345 (.105 + .240) she has one female chick; if it is between .345 and .670 (.345 + .325) she has two, and so on.

A sample program, corresponding to the model in section 8, might look like

Put in an initial population.
  Input C(0),J(0),A(0)

This calculates the number of female chicks from each adult female, and adds them up for the next generation of chicks.
    For $K = 1$ to (Number of Generations)
        For $i = 1$ to $A(K-1)$
            b =random #
            if $b \leq .05$, then $e = 0$
            if $.105 < b \leq .345$, then $e = 1$
            if $.345 < b \leq .67$, then $e = 2$
            if $.67 < b \leq .901$, then $e = 3$
            if $.901 < b \leq .985$, then $e = 4$
            if $.985 < b \leq .999$, then $e = 5$
            if $.999 < b$, then $e = 6$
            $C(K) = C(K) + e$
        next i
This counts how many chicks survive to be the next generation of juveniles.
        For $i = 1$ to $C(K-1)$
            s =random #
            if $s < .8$ then $J(K) = J(K) + 1$
        next i
This counts how many juveniles survive, taking into account the nonlinear survival rate.
        For $i = 1$ to $J(K-1)$
            s =random #
            if $s < .4(1 - \frac{A(K-1)}{500})$ then $A(K) = A(K) + 1$
        next i
This is the survival of the adults.
        For $i = 1$ to $A(K-1)$
            s =random #
            if $s < .7$ then $A(K) = A(K) + 1$
        next i
        Print K, C(K), J(K), A(K)
    Next K

10.3. Write a program to implement the above scheme.

10.4. Take the data from a 10-year run of the program you wrote in Exercise 9.3. Fit the model parameters from your data. How well did they agree with what was used in the model? Compare your results to other groups. Why is there a difference?

Since each time we run this program we get different values, due to the randomness in the program, this is not a solution in the same sense as we had before. We can, however, run the program many times and get statistics on the behavior. We can examine questions such as, "What happens if the available habitat is cut in half? What percent of the time will the population die out?" This kind of information is useful for wildlife managers who need to predict the effect of various actions. With endangered species, particularly, we can't run several experiments, so this sort of probability of survival is the best we can get.

10.5. Run your program for 100 years, starting with a group of 10 adults, 500 times. Get the following statistics:

    (a) Probability of survival for 100 years.

    (b) Mean and variance of average total population over the last 10 years.

    (c) Mean and variance of average adult population over the last 10 years.

Another possibility with a stochastic model is to add effects due to weather or epidemics, which some years have a much larger affect than others. A "weather multiplier" between 0 and 1 might be multiplied on all the survival rates, and the distribution of harsh winters estimated from past weather data. Similarly, a disease might affect survival rates by a lot in years of epidemic, and not at all other years. This can also be included in a stochastic model given sufficient data.

10.6. Suppose a frost 2 weeks later than average reduces egg production by 20%, and 4 weeks later than usual reduces eggs by 60%. Look up last frost dates in your area for the last hundred years and add this weather effect to your model. How are your statistics changed?

10.7. Suppose (on average) once every 20 years, pneumonia kills $\frac{1}{2}$ of all your birds. Add this effect to your model. How are your statistics changed?

11   CONCLUSION

We have gone through several methods to model population growth in our bird population. We started with linear models, getting the ideas of a growth rate and a stable age distribution. Then we looked at several possible nonlinear effects, and their fixed points. Finally, we considered how random effects could be considered.

This sort of modeling can be used to predict how changes in the environment affect populations. This can help with wildlife management. We have looked at fairly simple models, with birth and annual survival rates. Where necessary, more complex models can be built that consider more life stages: birth, fledging, juvenile survival, finding a territory, finding a mate, etc.

How good our predictions are is dependent on how good our data are. Another way these models are used is in helping to determine what information is most critical, and thus guiding what field work needs to be done. This is another important way that mathematics and science interact.

Of course, we could have used the same techniques for species other than birds. All we need is an understanding of the life cycle of the species, and the data.

Environmental Economics

Ginger Holmes Rowell
Middle Tennessee State University

INTRODUCTION

Water is necessary for sustaining life on Earth. Yet people are not always careful with this precious resource. People drink water to quench their thirst, eat fish from the rivers and oceans, use water to generate power, and enjoy water for fun and recreation. In fact, many people take for granted the seemingly unending supply of clean safe water. Yet, beaches are closed due to contaminated waters. Oceans are polluted with plastics that can kill wildlife, fish, and birds. Factories pump chemicals and sludge into waterways, causing contamination. This unit analyzes environmental issues involved with protecting water-related resources from an economical viewpoint.

Analyzing the costs and benefits of environmental issues is a complex task. However, such analysis is now a part of public policy. Executive Order 12866 requires that every regulatory ruling from the Environmental Protection Agency (EPA) of the Federal government must include a cost-benefit analysis [10]. According to the EPA, final decisions are based on public policy and not strictly on the cost-benefit analysis. The federal government recognizes the importance of using reliable methods to make economically sound decisions about protecting the environment. In this chapter, we assume the role of novice environmental economists in order to analyze the costs and benefits associated with various environmental issues. Many real-life environmental situations are complex. This unit uses a simplified perspective to learn some tools and techniques that will help us to make educated decisions about the future of the environment and the price of cleaning up environmental damage. We also introduce some of the difficulties that occur when applying these economic tools and techniques to complex environmental issues.

In particular, this unit uses hypothetical scenarios[1] to examine three environmental economics problems:

- Measuring the benefits of protecting the environment.

- Finding acceptable levels of pollution.

- Conserving common resources against overuse due to unrestricted access.

MEASURING THE BENEFIT OF PROTECTING THE ENVIRONMENT: THE TRAVEL COST METHOD

How much is it worth to protect parks, forests, mountains, rivers, and beaches? What is the value of fishing, picnicking, or just walking on a pristine beach? The value of such recreational areas where there is not a fee for usage is not easily measured. The **travel cost method (TCM)** is one way to place a monetary value on such areas.

> *The travel cost method determines an average amount of money that people are willing to spend in order to travel to a given area.*

A demand curve is determined by relating the cost of travel and the amount of use a location receives. The demand curve is used to measure **consumer surplus**.

> *Consumer surplus is the difference between what one is willing to pay and what one actually has to pay for a good or service.*

Economists measure the benefit of protecting natural areas by the increase that occurs in consumer surplus. The TCM attempts to evaluate the value of recreational areas by examining the willingness of travelers to spend money to travel to the area. The TCM can be used to help determine a fee price that will generate substantial revenue, to predict how management decisions could affect an area's value, or to measure environmental damage by comparing the area's value before and after environmental damage occurs.

Difficulties arise when trying to develop a method of placing a monetary value on nonmarket goods. Three issues of the TCM, which complicate determining the actual cost per trip [8], are as follows:

- Deciding what costs to include

- Determining the *opportunity cost* of the travelers' time

- Deciding how to correctly divide the cost of travel when the trip has more than one purpose or destination.

---

[1]The problems in these scenarios are variations of ideas presented in Goodstein's *Economics and the Environment* [5].

Scenario

Consider a fictitious beach community, Oceanfront City. For many years Oceanfront City has been a very popular tourist area. The area is growing and several large companies would like to build new businesses including power plants and chemical companies. The residents are worried that the pollution from the plants will contaminate the waters and kill fish, pollute the air with smog, and detract from the overall quality of their tourist region. However, they also like the promise of additional jobs and revenue brought into the community. The community task force is using several kinds of economic analysis to help decide if the businesses should be brought into the community. The TCM is chosen as one means of placing a monetary value on the environment that people are currently using for free. The task force wants to know how much people are willing to spend in travel costs to get to Oceanfront City. Of course, many other factors are considered in the task force's final decision.

Using the following simplified example, we will learn how to use the TCM. Begin by thinking of placing concentric circles around Oceanfront City. Classify the area between the circles as separate regions. Assume that within a given region the cost to travel to Oceanfront City is approximately the same. Generally, the population within the region closest to the destination area will make more trips to the area and have a lower travel cost per trip. See Figure 1 below.

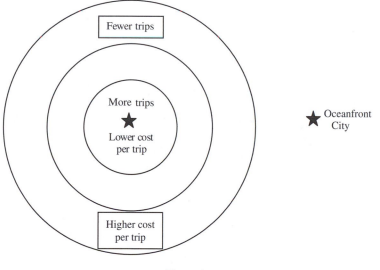

Figure 1.

The Oceanfront City community task force took an in-depth survey of tourists already visiting their area. For simplification, we will assume the influences of age, income, and other factors are not included in the data shown below.

Number of Trips to Oceanfront City per Season	Travel Cost to Oceanfront City per Trip	Total Travel Cost Per Season
30	$ 9	$270
15	$ 27	$405
7	$ 54	$378
3	$100	$300

Let's begin by inspecting the data to make certain that we understand each component. According to the data, the travelers who make an average of 30 trips per season to Oceanfront City are willing to spend an average of $9 per trip for a total travel cost of $270. Equivalently, we could say that the travelers who spend an average of $9 per trip make an average of 30 trips to Oceanfront City during the season. Verify the total travel cost column. Notice that the total cost is not necessarily an increasing function of the number of trips because of the decreasing cost associated with being in a region nearer the area.

Now let's examine the data graphically. Begin by drawing a concentric circle diagram to represent the data. Recall that this type of diagram assumes that the cost within a given region is approximately the same. Compare your graph to the one in Figure 2.

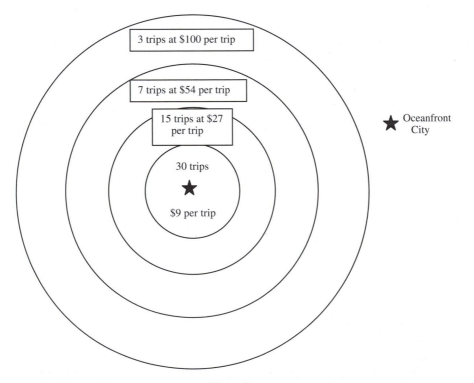

Figure 2.

Since our goal is to put a monetary value on the recreational value of Oceanfront City, let the cost per trip represent the output variable. Make a scatter plot of the cost per trip for a family to travel to Oceanfront City as a function of the number of trips made to Oceanfront City per season. Label your axes. Title your graph. Compare your graph with Figure 3 below.

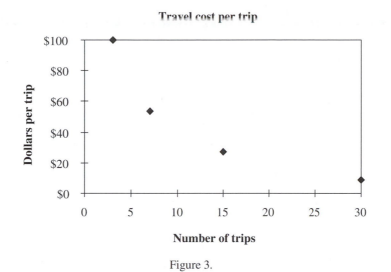

Figure 3.

The curve that passes through these points represents the demand (in dollars per trip) for trips to Oceanfront City. The curve connecting the data points appears to have an exponential or hyperbolic shape. In this example, we find a linear estimate of the demand curve and leave it as an exercise to re-examine this problem using other functions. You can use the linear regression feature of a graphing calculator to determine the line that best represents the data points. This example demonstrates how to complete this problem without the use of a graphing calculator. We can find a linear estimate to the data by placing an imaginary black thread in a straight line between the data points and adjusting this thread until the vertical distance from the points to the thread is as small as possible. Select any two points on the line of the black thread to use to find the equation of the line. One possible linear estimate of the data would occur if the black thread passes through points (5, 73) and (25, 13) as is shown in Figure 4 below. (Students' imaginary black threads may be placed in slightly different locations.)

One method for finding the equation of this line is to first find the slope and then use the slope-intercept form. The formula for slope is:

$$\text{slope} = m = \frac{\Delta y}{\Delta x} = \frac{y_2 - y_1}{x_2 - x_1}$$

Since we chose to represent the variables for this problem as number of trips on the horizontal or x-axis and dollars per trip on the vertical axis or y-axis, then the slope is found

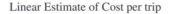

Linear Estimate of Cost per trip

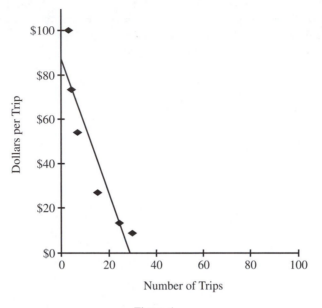

Figure 4.

by

$$\text{slope} = \text{change in dollars spent per trip/change in number of trips}$$

Using the two chosen points $(5, 73)$ and $(25, 13)$, we find the slope as follows:

$$\text{slope} = m = \frac{73 - 13}{5 - 25} = \frac{60}{-20} = -3$$

Comparing this answer with the scatter plot verifies that the slope should be negative. The demand curve is a decreasing function of the number of trips made. This seems reasonable. If all other factors are equal, the travelers from the closer regions will have lower costs per trip and make more trips.

Next, we find the equation of the line. This can be done using the slope intercept form, $y = mx + b$, or the point-slope form, $y - y_1 = m(x - x_1)$. We will demonstrate the point-slope method. Choose any point on the black thread line as (x_1, y_1); we will use $(5, 73)$. Substitute $(x_1, y_1) = (5, 73)$ and the slope $m = -3$ into the point-slope form to find the equation of the line passing through these points.

$$y - y_1 = m(x - x_1)$$
$$y - 73 = -3(x - 5)$$
$$y - 73 = -3x + 15$$

$$y = -3x + 15 + 73$$

$$y = -3x + 88$$

The equation is now in the form $y = mx + b$, where m is the slope and b is the y-*intercept*. Notice on the scatter plot, if we extend the demand curve to cross the y-axis it will cross at approximately 88. This is a good check for our computations. We can write our equation as a demand function

$$D(x) = -3x + 88,$$

where x is the number of trips to Oceanfront City per season, and demand represents an average dollar amount per trip that consumers are willing to pay for travel.

Based on demographic information of the areas surrounding Oceanfront City, the task force has estimated that 50,000 visitors are willing to make an average of 10 trips per season to Oceanfront City. Using this information, find the average cost of a trip to Oceanfront City.

Using our demand equation $D(x) = -3x + 88$ and letting $x = 10$ (the average number of trips) we can find the average travel cost per trip.

$$D(x) = -3x + 88$$

$$D(10) = -3(10) + 88$$

$$= 58$$

The average visitor to Oceanfront City is willing to spend $58 *per trip* in travel costs.

Graph this constant average demand function, Avg $D(x) = 58, on the same axes with the linear estimate for the original data, as shown in Figure 5.

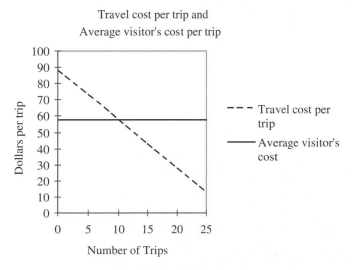

Figure 5.

Notice that if the task force had told us that the average visitor is willing to spend $58 per trip in travel costs to get to Oceanfront City, then we could determine the number of trips that the average person is willing to make to Oceanfront City. Using the demand equation found from our estimate of the data, $D(x) = -3x + 88$, we would set $D(x) = \$58$ and solve for x. The work below shows that if $D(x) = \$58$, then the average travelers make $x = 10$ trips per season.

$$D(x) = -3x + 88$$
$$58 = -3x + 88$$
$$58 - 88 = -3x$$
$$-30 = -3x$$
$$10 = x.$$

Now we will use this information to put a value on the environment of Oceanfront City. One way economists measure the value of protecting environmental areas is the increase in **consumer surplus**.

> *Consumer surplus is the difference between what one is willing to pay and what one actually has to pay for a good or service.*

For example, on a really hot day, you might be *willing* to pay $2 for an ice-cream cone, even if the ice-cream cone only costs $1.50. Your willingness to pay $.50 above the actual cost is the *consumer surplus*.

To find the consumer surplus for the average visitor to make a single visit to Oceanfront City, we need to compare the amount one is willing to spend to make one trip with the actual cost of one trip for the average traveler. The amount one is willing to spend for one trip to Oceanfront City is found by evaluating the demand function $D(x)$ at $x = 1$ trip.

$$D(x) = -3x + 88$$
$$D(1) = -3(1) + 88$$
$$= 85.$$

Thus one is willing to pay $85 in travel costs for one trip to Oceanfront City. However, we determined that the average visitor only spends $58 in travel costs per trip. The consumer surplus for the average visitor to make one trip to Oceanfront City is found by

Consumer surplus = amount willing to pay − amount actually have to pay

$$= \$85 - \$58$$
$$= \$27.$$

Now we want to find the **total consumer surplus** for the average visitor. The total consumer surplus is easily visualized using a graph.

> *The consumer's surplus (also called the* consumer's net benefit*) is the area under the consumer's demand curve minus the area representing total consumer cost.*

Graphically, the total cost is equal to the area under the marginal cost curve. Marginal cost is the extra cost per extra trip. Total cost is the cost per trip multiplied by the number of trips. Since it costs the average visitor $58 per trip, and the average visitor makes 10 trips to Oceanfront City, then the total cost for those 10 trips is $(10) \times (\$58) = \$580$. This is equal to the area under the average cost function shown below. The area under the demand curve represents the consumer's total willingness to pay or the total demand and forms a trapezoidal region. The consumer surplus is the shaded triangular region in Figure 6.

Consumer surplus

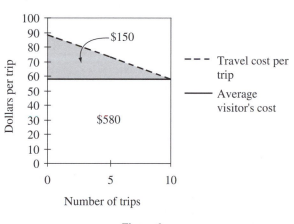

Figure 6.

Consumer surplus = (area under demand curve) − (area representing total consumer cost)

$$= \text{area of trapezoid} - \text{area of rectangle}$$

$$= \text{area of the triangular shaded region}$$

$$= (1/2)bh$$

$$= (1/2)(10)(88 - 58)$$

$$= 150$$

The total consumer surplus for the average visitor in one season is $150.

The **expected consumer surplus** for the season is the total season consumer surplus for the average visitor multiplied by the anticipated total number of visitors for that season.

Expected Consumer Surplus = (total consumer surplus per visitor) × (number of visitors)

$$= (\$150 \text{ per visitor})(50{,}000 \text{ visitors})$$

$$= \$7{,}500{,}000$$

So what would you recommend to the Oceanfront City task force? Should the industrial business plants and factories be built? If these data are accurate, how much revenue would the new businesses have to generate for Oceanfront City in order to make the plant production worthwhile?

THE ACCEPTABLE EFFICIENCY STANDARD FOR POLLUTION: MARGINAL ANALYSIS

How much pollution is too much? Ideally, one might say that all pollution should be eliminated. In reality, it is too expensive to eliminate all pollution. It is more realistic to establish an *efficient* level of pollution, and work toward the goal of achieving that level. Economists define *efficient* as the condition where it is impossible to make one person better off without making anyone else worse off. Environmentalists want to know if the levels of pollution reduction that are considered efficient levels for today's economy will remain efficient or safe in the future.

When the objective is to reduce pollution one unit at a time, then **marginal analysis** can be used to determine efficient levels of pollution reduction. **Marginal costs** represent the change in the total cost function, or the cost to produce one additional unit. **Marginal benefits** represent the change in the total benefits.

The level of pollution that is the most "efficient" is the level of pollution where the marginal costs equal the marginal benefits.

We will use marginal analysis to study the economics of cleaning up polluted beaches.

Almost 4000 beaches were closed or issued swimming advisories by state and local governments in 1995 [3]. Harmful microorganisms typically found in improperly treated sewage cause a pollution hazard for beaches. Sewage can enter beach waters from sewer overflows, malfunctioning sewage treatment plants, illegal dumping, and untreated storm water runoff. Due to the yearly numbers of beach closures increasing, the EPA initiated a program to help protect the public health at our beaches. In addition to personal health risks, polluted beaches can cost a community money by discouraging tourists.

How much is it worth to clean up polluted beaches or other polluted areas, and how do we measure the benefits? The city of San Francisco, California decided that controlling sewer overflow was worth $1.45 billion and twenty years of work to create what the EPA has recognized as the best combined sewer overflow control system in the nation. Before the control system was in place, San Francisco had 50 to 80 untreated shoreline overflows a year. Now they only have 1 to 10 shoreline overflows per year [4].

Measuring benefits from clean beaches is more than just determining the gain in revenue. If the pollution is a health hazard, then avoiding health care costs is considered a benefit. This type of economic evaluation becomes even more complicated when one tries to measure the value of human life and suffering. In this project, we will study simplified cost-benefit analysis.

One measure of the value of clean beaches to people is the proliferation of beach cleanup days and the increasing number of volunteers. Beach cleanup projects began in 1984 in Oregon with 2100 people and grew to about 100,000 volunteers throughout the coastal regions in 1990 [7].

Scenario

Consider again the beach community, Oceanfront City. The city council has been concerned about the growing levels of pollution in the ocean. On several occasions the beach was closed due to potential health hazards caused by pollution. Tourism, a large source of revenue for the community, is declining. The city council is asking the residents to volunteer their time to help clean up the beach areas. Additionally, the city is committed to upgrading the sewage system. Additional revenue is needed to pay for these beach cleanup plans. Currently visitors do not pay a fee to enjoy the public beaches at Oceanfront City. The city planners are trying to determine the effects of a fee system on the economy. They are discussing charging for parking, having a weekly admission charge like the national park system, and season passes along with other options. Residents and tourists of Oceanfront City were given a survey to help the planners determine the economic value of reducing the pollution in the beach community. The surveyors classified the amount of pollution reduction as levels of cleanup and clearly defined and explained these levels in the survey. The survey responses were averaged and the results are shown below.

Amount of Pollution Reduction	Beach Fee Average Resident is Willing to Pay *After* Cleanup
none (Level 0)	$ 0
very small amount (Level 1)	$ 4
small amount (Level 2)	$ 7.5
medium amount (Level 3)	$ 9
large amount (Level 4)	$10
very large amount (Level 5)	$10.50

The average amount per person that the visitors are willing to pay is called the total benefit for one individual. Assume, as in the travel cost example, that there is a population base of 50,000 visitors willing to make 10 visits for the current rate of beach cleanliness if no fee is charged. Therefore the expected number of visits to Oceanfront City is approximately 500,000 visits per season. The total *market* benefit for Oceanfront City will be the value for the average *individual* benefit multiplied by the total number of visits. For the first level of cleanup the total market benefit is $4 \times 500,000 = $2,000,000$. Complete the chart below.

Level of Cleanup	Individual Total Benefit	Market Total Benefit
0	$0	$0
1	$4	$2,000,000
2	$7.50	
3	$9	
4	$10	
5	$10.50	

First we will examine the marginal benefits gained by Oceanfront City from cleaning up the beach. **Marginal benefits** are the benefits gained for reducing one additional unit of pollution. In our scenario, the level of cleanup measures the pollution reduction, and the benefit gained for an extra level of cleanup is the marginal benefit. Determine the marginal benefit by taking the difference between the total benefits for completing one additional level of cleanup. For example, assume we start with no cleanup and no fee for using the beach. For the first level of cleanup, the survey indicates that people are willing to pay $4. Thus, $4 is the marginal benefit for the first level of cleanup. After the second level of cleanup people are willing to pay a total of $7.50. This amount is $3.50 more than they are willing to pay for the first level of cleanup. Thus the amount they are willing to pay for that second additional level of cleanup, $3.50, is the marginal benefit. Verify the results in the *individual* marginal benefits column and complete the marginal benefit for the entire *market* of Oceanfront City.

Level of Cleanup	Individual Marginal Benefit	Market Marginal Benefit
0	—	—
1	$4	
2	$3.50	
3	$1.50	
4	$1	
5	$.50	

Use a scatter plot to gain a visual representation of the data points. Since we want to measure the benefits, then choose benefit as the output variable on the y-axis and choose the level of cleanup as the input variable on the x-axis. Label the axes and plot the data points. Compare your graph to Figure 7. Notice the relationship between these points. The relationship is approximately linear. As described in the travel cost example, use the black thread method or a linear regression tool to find the equation of a straight line estimating the data. Verify that a linear approximation of the data is

$$y = -.95x + 4.95.$$

Marginal benefits for beach cleanup

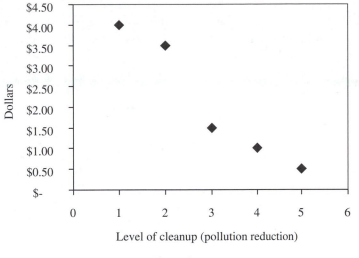

Figure 7.

Thus the marginal benefit function, *MB*, for the average individual in Oceanfront City is

$$MB(x) = -.95x + 4.95.$$

Compare your result with the marginal benefit equation above and the graph below. Since these are approximations the answers may vary slightly. Figure 8 provides a graphical representation of the linear estimate of the marginal benefit of cleanup.

Marginal benefits for beach cleanup

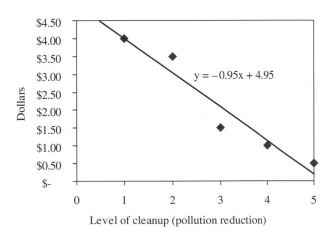

Figure 8.

Next we will examine marginal cost. The **marginal cost** is the cost that is incurred from cleaning up one additional unit. The city council has received cost estimates for various levels of upgrades for the sewer system, for disposing of pollution collected by volunteers, and for other cleanup costs. Upgrading the sewer system is an expensive and time-consuming project and the city council has devised a five-year plan. The city council felt the users of the beach, both residents and tourists, should share the cost of reducing pollution. Dividing the total cost for the pollution reduction by the expected number of visitors and the number of years provides an estimate of the individual cost per season to reduce pollution on the beaches. The council was given the following results.

Level of Cleanup	Individual Total Cost per Season
0	$0
1	$1.50
2	$3.00
3	$4.50
4	$6.00
5	$7.50

Use this data and the example of the marginal benefits to complete the marginal costs column in the following chart.

Units of Cleanup	Individual Marginal Cost
0	—
1	$1.50
2	
3	
4	
5	

From the data we conclude that the *individual* marginal cost function is a constant function for $x = 1, 2, 3, 4, 5$ levels of cleanup

$$MC(x) = \$1.50.$$

Now we are ready to find the **efficient level of pollution reduction**.

The efficient level of pollution reduction occurs where marginal benefits equal marginal costs.

Graphically, the point of intersection between the individual marginal cost and individual marginal benefit indicates the efficient level of pollution. Graph the marginal cost function and the marginal benefit function on the same axis and compare your graph with Figure 9.

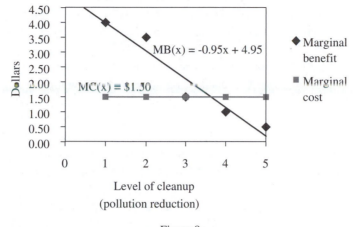

Figure 9.

The graph may provide you with only an approximate solution depending on the resolution of the graph and the values of the solutions. Since the marginal cost and marginal benefit equations are linear, two algebraic methods of **substitution** and **elimination** can be used to find a solution. Use one of these methods to find the level of cleanup, x, that produces an efficient level of pollution reduction. According to the cost-benefit analysis, at what level should the city council pay to clean up the beach? Since the solution for x is not an integer value, we examine the graph and see that Level 3 is the efficient level of cleanup. The graph shows that the residents are not willing to pay for the fourth level of cleanup since the marginal costs are higher than the marginal benefits for that unit. Do you agree with the results? What other factors should be considered? What are the limitations of this type of analysis? To find out what you can do to help prevent beach pollution see the references in the exercises.

CONSERVATION OF COMMON RESOURCES: THE PROBLEM OF FREE ACCESS

Some resources are not privately owned, but are considered common resources. Giving people free access to common resources generally results in exploitation of the resources. Exploitation occurs when people weigh the private benefit against the private cost instead of against the social cost. Damage to these resources can occur at a level that is inefficient or unsafe. We examine the environmental concern of overfishing that occurs when free access is given to the common resource of oceans and other waterways.

A very real catastrophe demonstrating the devastating consequences of overfishing occurred in Canada's East Coast fishery in Newfoundland. Canadian fishermen voluntarily stopped fishing for cod off the Grand Banks of Newfoundland in February 1992, because the fish were too small. The fishermen could not make a profit despite their high-tech equipment. In July 1992, approximately 25,000 fishermen were out of work when New-foundland's entire in-shore fishery was closed. The Canadian government had to spend millions of dollars to retrain fishermen for other jobs [9]. Could this catastrophe have been prevented? Will the cod population of Canada's East Coast rebuild itself? If so, how long will it take?

Canada's overfishing problem is just one of many such problems around the world. According to the United States Food and Agriculture Organization, 69% of the world's fisheries have been heavily exploited [6]. *Overfishing* can be divided into two categories: *economic overfishing* and *biological overfishing*. Economic overfishing occurs in unregu-lated fisheries because of exploitation of common resources and causes inefficient use of resources. Biological overfishing reduces the fish stock to a level at which the productivity declines [2]. The National Oceanic and Atmospheric Administration (NOAA) has a plan to address these over-harvesting issues.

NOAA is committed to greatly increasing the nation's wealth and quality of life through a healthy fishing industry. To achieve this goal, NOAA is working to ensure *sus-tainable* fisheries that provide safe and wholesome seafood and recreational opportuni-ties [11]. The concept of *sustainability* can be defined in a broader environmental sense.

> *Sustainability of resources insures that future generations will have comparable materials and environmental welfare as we do now.*

A sustainable yield of fish can be generated when the harvest rate does not exceed the growth rate of the fish supply. If fish and whales are killed at a rate faster than they can reproduce, then their species will become extinct. The underlying principal for regulating fisheries is the *maximum sustainable yield*.

> *The maximum sustainable yield (MSY) is the highest possible rate of use for which a system can replenish on its own.*

Fishery managers use MSY estimates to set *total allowable catch* (TAC) limits. Determin-ing the MSY has many complicating factors including inadequate data and unusual weather occurrences like hurricanes. We will examine a few of the many concepts of the economics of managing fisheries in the following scenario.

Scenario

Fishing is one of the main industries for Oceanfront City. Like other fishing communi-ties, over time Oceanfront City has seen a decline in the area's total fish catch despite the fact that more boats have been fishing the area. The local fishing commission wants to understand the problem and has invited professors from a local college to help citizens understand the economics of managing their fishing industry and to find a solution to the problem before it gets worse.

To better understand the economic aspects, the commission first explored the biologi-cal aspects of fishing populations. The population dynamics of a fishing population undis-

Population

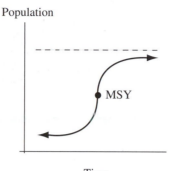

Time

Figure 10.

turbed by man is demonstrated in Figure 10 (called a *sigmoid curve*). Over time, we can
see that the environment has a natural limit to the size of population that can be supported.
The maximum sustainable yield (MSY) is the point on the graph where the fastest rate of
increase in the fish population occurs.

Next the commission learned that according to the simplified Shaefer model the quantity of fish caught as a function of the effort used to fish is parabolic [1]. In order to understand the concepts before working with actual data for the Oceanfront City fishing industry,
the commission was given the simplified collection of information and data below, which
is a variation of a problem by Goodstien [5].

The daily fishing cost per boat is $200. This includes salaries for the crew, fuel and
other operating costs. For ease of computations, we will assume that the fishermen sell the
fish for $1 per pound.

Number of Fishing Boats	Total Catch per day (pounds)
2	1600
3	2100
4	2400
5	2500
6	2400
7	2100
8	1600

First examine the data and its implications. If three boats were sent out each day for the
entire season, then the expected total daily catch would be 2100 pounds per day, which
is 700 pounds per boat per day. We would expect for the total catch to increase as the
number of boats increases. However, as more boats catch more fish, over time the fish
population will decrease. Since there are less fish available to catch, the daily catch totals
decrease even though more boats are fishing. This data represents three important concepts
in fishery management: expansion, MSY, and overfishing. Can you determine an estimate
of when each of these concepts occurs? Do you agree with the following chart?

Number of Fishing Boats		Total Catch per day (pounds)
2		1600
3	expansion	2100
4		2400
5	MSY	2500
6		2400
7	overfishing	2100
8		1600

The fishermen will **break even** if the money they make equals the money they spend. The money made from selling the fish is called **revenue**. The money spent on one day of fishing is called the daily **cost**.

A fishing boat will break even if its revenue equals its cost.

Since the selling price for fish is $1 per pound, the total daily revenue equals to the total daily catch. Using the example data provided, we can find the **average catch per boat** by dividing the total daily catch by the number of boats fishing that day. Verify and complete the results in the average catch column of the following table.

Number of Fishing Boats	Total Catch per day (pounds)	Average Catch (pounds) per Boat per day
2	1600	800
3	2100	700
4	2400	
5	2500	
6	2400	
7	2100	
8	1600	
9	900	

Assuming that the fishermen will want to at least break even, they will not take their boats out if their average revenue is less than the assumed cost of $200 per day. Examine the chart to find the break-even point. According to the data, only eight boats will go out per day, since the daily cost per boat equals the average daily revenue per boat for eight boats. Therefore, the fishers of Oceanfront City will regulate themselves and take out on average a total of only 8 boats per day.

However, the fishing commission is interested in the good of entire community and wants to determine the number of boats that should be allowed to fish in the bay in order to maximize the **total profit** for the communities fishing industry.

The total profit is the total revenue minus the total cost.

The total revenue is easy to determine since the selling price for the fish is $1 per pound. The total cost for the fishing industry is the cost per boat times the number of boats. The following chart can be used to help determine the number of boats that will maximize profits. The total profit column is calculated by subtracting the total cost from the total revenue for each number of boats.

Number of Boats	Total Revenue for the Industry	Total Costs for the Industry	Total Profit for the Industry
2	$1600	$ 400	$1200
3	$2100	$ 600	$1500
4	$2400	$ 800	$1600
5	$2500	$1000	$1500
6	$2400	$1200	$1200
7	$2100	$1400	$ 700
8	$1600	$1600	$ 0

In order to maximize the total profits from the fishing industry, would you restrict free access to the fishery? If so, examine the total profit column to determine the number of boats that you would allow to fish per day. Did you conclude that four boats per day would maximize profits for the fishing area? How does this compare with your earlier estimates for MSY? If the commission limits the number of boats to provide the maximum profits for the fishing industry, will the fishery remain sustainable for future generations?

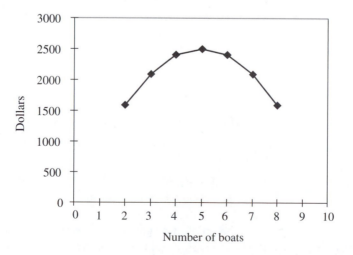

Total revenue for the bay per day

Figure 11.

Since realistic data is not nearly as simplified as the commission's hypothetical data, we also need to study an algebraic method of analysis. Use a scatter plot to graphically represent the total revenue as a function of the number of boats. The curve that models the data is shown in Figure 11. From the graph, we see that the total revenue from the fishing industry has a parabolic shape as was expected (according to the simplified Shaefer model [1]), since the number of boats represents effort expended. In general, you could examine the points and the vertex to find a quadratic equation for total revenue. A calculator or computer program that computes quadratic regression coefficients will also help you determine a functional representation for the data. Let x represent the number of boats per day and write $R(x)$ to be the total revenue function. Then

$$R(x) = -100x^2 + 1000x.$$

The total cost function is the cost per boat, \$200, times the number of boats, x.

$$C(x) = \$200x.$$

Total profit function is the total revenue function minus the total cost function.

$$P(x) = -100x^2 + 800x.$$

The profit function is a quadratic function. Thus its graph is a parabola.

For $y = ax^2 + bx + cy$, if $a < 0$, then the parabola opens downward, resulting at a maximum value at its vertex.

The x coordinate of the **vertex** can be found by the formula

$$x = -b/(2a).$$

Then the corresponding function value or y-value is found by substituting the answer for x into the original function. For the given profit function, $a = -100$, $b = 800$, and $c = 0$. Since $a < 0$, the parabola opens downward. Thus, the profit is maximized at its vertex. Find the vertex of the profit function's graph.

$$x = -b/(2a) = -800/((2)(-100)) = 4.$$
$$y = P(4) = -100(4^2) + (800)(4) = 1600.$$
$$\text{vertex} = (x, y) = (4, \$1600)$$

Using this algebraic approach we confirm the result of our table method and conclude that four boats will maximize the fishing industry's profit, and the maximum profit is \$1600 per day.

Now the question arises of what the fishing commission should do. Without regulations, eight fishing boats will go out per day and the fishing industry will not maximize its profits. More importantly, the eight boat limit is substantially above the MSY limit of five boats and could cause a depletion of the fish stock. Some ecologists question whether recovery is always possible [6].

As in the previous section, to find the economically **efficient** level of effort (the numbers of boats) set marginal cost equal to marginal benefit (revenue) and solve for x. Thus we need to find the marginal cost and marginal revenue curves. The marginal cost is the cost per day and is a constant of $200 per day, $MC = 200$. The marginal revenue function can be found by taking the differences in the total revenue data. Verify the marginal revenue column of the chart below.

Number of Boats	Total Revenue	Marginal Revenue
2	$1600	—
3	$2100	$500
4	$2400	$300
5	$2500	$100
6	$2400	−$100
7	$2100	−$300
8	$1600	−$500

To find the equation of the marginal revenue function, select any two coordinate points of the form (number of boats, marginal revenue) and use the point-slope form. Verify that

$$MR(x) = -200x + 1100.$$

To find the efficient level of effort, set the marginal cost equal the marginal revenue and solve for x.

$$MR(x) = MC(x)$$

$$-200x + 1100 = 200$$

$$-200x = -900$$

$$x = 900/200$$

$$x = 4.5 \text{ boats}$$

Since we will not send out one half of a boat, we must select the correct number of boats to send out. Since the limit for MSY is five boats, then the commission is likely to select five boats. However, if five boats were sent out daily, then the marginal cost would exceed the marginal revenue, and this is not the most efficient level. In real life situations, many factors must be weighed in making such decisions including the employment of citizens and the sustainability of the fishery. What would you do?

EXERCISES

Travel Cost Method (TCM) Supplementary Exercises

1. Explain the goal of the TCM.

2. Explain the weaknesses or limitations of the TCM.

3. Assume Oceanfront City received the following results from their travel cost survey.

Number of Trips to Oceanfront City per Season	Travel Cost to Oceanfront City per Trip
30	$7
20	$15
10	$40
5	$60

(a) Find the total cost for each number of trips.

(b) Sketch a scatter plot of the given data with the travel cost per trip as the output variable.

(c) Find a linear approximation of the data and write it as a demand function for trips.

(d) Based on demographic information, assume the population base supports 35,000 visitors willing to make an average of 12 trips per season. Find the average travel cost per trip.

(e) Graph the constant average demand function found in part (d) for the average traveler on the same graph with the demand function found in part (c).

(f) Find the total consumer surplus for the average visitor for one season.

(g) Find the total expected consumer surplus for all visitors for the season.

4. The TCM has been used as a means of estimating the economic value of recreational salmon fishing on the Gulkana River in Alaska under the current conditions and under hypothetical fishing management conditions [8]. The travel cost to Gulkana ranges from 0 to $1,573, with an average cost of $418. The average number of fishing trips to Gulkana per season is 2.67. Therefore the point (2.67, $418 per trip) is on the TCM demand curve.

(a) If the travelers making two trips are willing to pay $460 per trip, find the demand curve for trips to Gulkana. (Assume that a *linear* function is appropriate for modeling the demand.)

(b) Graph the demand curve and the horizontal line representing the average cost of $418.

(c) Shade the region that represents the average individual's total consumer surplus.

(d) Find the numerical value of the average individual's total consumer surplus.

(e) What is the total expected consumer surplus for the region if 5,000 visitors are expected during the season?

5. (Calculator Problem) Refer to the data below for Oceanfront City and use a graphing calculator with a regression feature to express the demand function as an exponential model of the data. Is this a better model than the linear model in the example? Explain your reasoning.[2]

Number of trips to Oceanfront City per year	Travel Cost to Oceanfront City per Trip	Total Travel Cost for all Trips
30	$9	$270
15	$27	$405
7	$54	$378
3	$100	$300

Reducing Pollution: Cost-Benefit Analysis Supplementary Exercises

1. Explain what an efficient level of pollution means.

2. Define **marginal costs** and **marginal benefits**. Explain how to use these concepts to complete cost-benefit analysis.

3. Explain the weaknesses or limitations of cost-benefit analysis.

4. For the given example of Oceanfront City, use **substitution** or **elimination** to find the level of cleanup, x, to verify the graphical results that Level 3 is the **efficient** level of pollution reduction. Do you agree with the results? What other factors should be considered?

5. Assume that Oceanfront City was given the following survey results for reducing the pollution on the beach. Recall that Level 1 cleanup is a very small amount of pollution reduction and Level 5 is a very large amount of pollution reduction.

Level of Cleanup	Total Cost	Total Benefits
0	$0	$0
1	$4	$10
2	$9	$18
3	$15	$24
4	$22	$28
5	$30	$30

(a) Find the marginal costs and marginal benefits for each level of cleanup.

(b) Use a graphical method to represent the efficient level of pollution reduction.

(c) Determine the linear equations for the marginal costs and marginal benefits and use an algebraic method to verify the results found by the graphical method.

---

[2]Depending on the level of the class, the entire example problem could be revisited using this exponential model. Area approximations could be found for the consumer surplus.

6. What should Oceanfront City do if they received the following survey results?

Level of Cleanup Costs	Individual Marginal Costs	Individual Marginal Benefits
1	$ .50	$12.50
2	$ .75	$8.00
3	$1.00	$4.50
4	$1.25	$2.00
5	$1.50	$.50

Graph a scatter plot of the marginal cost and connect the points with a smooth curve. The relationship appears to be linear. Next, graph the scatter plot of the marginal benefits. These points do not fall in a straight line, but instead are parabolic in nature. A calculator that performs quadratic regression on your data (x = number of days of cleanup, and y = marginal benefits) can easily come up with the equation for this data. Verify the following marginal cost and benefit functions model the given data.

$$MC(x) = .25x + .25$$

$$MB(x) = .5x^2 - 6x + 18.$$

Find the efficient level of pollution reduction for these survey results.

Resources and Questions for Learning More about Preventing Beach Pollution

1. Explore the EPA's "Beachgoer's Guide" on the Internet at

http://www.epa.gov/ostwater/beaches/goer2.html

   (a) What is the EPA doing to make beaches safer?

   (b) What costs are associated with these improvements?

   (c) Define **nonpoint-source pollution**.

   (d) What can you, your family, and/or your school do to try to help protect surface and ground waters from nonpoint-source pollution? What types of costs are involved with making these improvements?

   (e) Continue exploring the EPA's "BEACH Watch" web site and create a report with graphs showing what is happening with beach closings.

           http://www.epa.gov/OST/beaches/index.html

Conservation of Common Resources: The Problem of Free Access

1. Define the following terms: **cost**, **revenue**, **profit**, **break-even point**.

2. What do the terms **marginal cost** and **marginal revenue** represent?

3. Explain the follow terms: **sustainability**, **sustainable yield**, and **maximum sustainable yield**.

4. On the graph in Figure 12, mark the point of maximum sustainable yield. Explain your answer.

Population

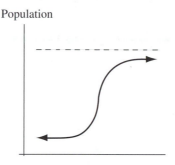

Time

Figure 12.

5. Assume that Oceanfront City received the following information about their fishing industry.

Number of Boats Fishing	Total Revenue	Total Cost
4	$1600	$800
5	$2000	$1000
6	$2300	$1200
7	$2000	$1400
8	$1600	$1400
9	$1000	$1800

(a) Examine the data by inspection.

(b) Find the point of MSY. Where is the fishing industry expanding? Where is over-fishing occurring?

(c) Determine how many boats should fish in order for the fishermen to break even.

(d) Determine the total profit for the fishing industry for each number of boats fishing.

(e) Graph the total profit as a function of the number of boats fishing.

(f) Find the efficient level of fishing.

(g) What do you recommend to the fishing commission?

REFERENCES

1. Frederick W. Bell, *Food From the Sea: The Economics and Politics of Ocean Fisheries*, Westview Press, Inc., Boulder, Colorado, 1978.

2. Colin W. Clark, *Bioeconomic Modeling and Fisheries Management*, John Wiley and Sons, New York, 1985.

3. Environmental Protection Agency, "US EPA's BEACH Watch: Frequently Asked Questions," http://www.epa.gov//OST/beaches/faq.html, 7/1999.

4. Environmental Protection Agency, "US EPA's BEACH Watch: Success Stories," http://www.epa.gov/OST/beaches/prevent.html/#sanfran, 6/1997.

5. Eban Goodstein, *Economics and the Environment*, Prentice-Hall, New Jersey, 1995.

6. Mike Hagler, "Deforestation of the Deep: Fishing and the State of the Ocean," *The Ecologist*, 25 No 2/3 (1995) 74–79.

7. Edward Laws, *Aquatic Pollution: An Introductory Test*, 2nd ed., John Wiley and Sons, Inc., New York, 1993.

8. Chris Layman, John R. Boyce, and Keith R. Criddle, "Economic Valuation of the Chinook Salmon Sport Fishery of the Gulkana River, Alaska, under Current and Alternate Management Plans," *Land Economics*, 72(1) (1996) 113–128.

9. David Ralph Matthews, "Commons versus Open Access: The Collapse of Canada's East Coast Fishery," *The Ecologist*, 25 No. 2/3 (1995) 86–96.

10. Bernard Nebel and Richard Wright, *Environmental Science: The Way the World Works*. 5th ed. Prentice Hall, New Jersey, 1996.

11. National Ocean and Atmospheric Administration, "Sustainable Fisheries," http://www.noaa.gov/sustainable_fisheries.html 7/1999.

12. Tim Tietenberg, *Environmental and Natural Resource Economics*. 3rd ed, Harper Collins, New York, 1992.

Some Mathematics of Oil Spills

Donald E. Miller
Joanne E. Snow
Saint Mary's College, IN

INTRODUCTION

Oil, sometimes called Black Gold, is currently the largest energy-producing natural resource in the world. It is found under the surface of the earth in several locations and must be removed and transported to locations where it is used or refined to form fuels such as gasoline. Much of this transportation is done through pipelines, but a large amount is on water in huge tankers with capacities of several million gallons. In fact, more than 400 billion gallons is transported by sea each year [15, p. 40]. With this method of transportation responsible for such volume, the loss of some oil into the water, an oil spill, is inevitable. It is this hazard of meeting the world's increasing need for energy we discuss in this module.

Oil spills can result from human error or from equipment failure. Whatever the cause, the loss of large quantities of oil into a body of water can cause serious ecological problems. The potential for human error is highest in the loading and unloading process. These tankers are loaded and unloaded by pumping oil through large hoses that resemble fire hoses but are much larger and stronger. Human operators must attach these hoses to the tanker and operate the pumps. If a hose comes loose during the pumping process, large quantities of oil can be dumped into the harbor in just a few minutes. Other possibilities include a hose breaking, an operator starting a pump too early, or turning it off too late, causing the tanker to overflow.

A large tanker is being filled with crude oil for transport from the fields in Alaska to a refinery. The oil is being pumped into a tanker at the rate of 400,000 gallons per hour. Use this scenario in answering Exercises 1–4.

191

Exercise 1. If the hose were to break away from the tanker and the operator took eight minutes to stop the pump, how much oil would be dumped into the harbor?

Exercise 2. If two pumps are used, how long would it take to fill a tanker with a capacity of 12 million gallons?

Exercise 3. At 7:00 P.M., the hose springs a leak which continues to enlarge until the operator notices it and shuts off the pump at 7:15 P.M. When the pump was stopped, it is estimated that 25% of the oil being pumped through the hose was leaking into the harbor. Estimate the amount of oil spilled into the harbor as the result of this equipment failure and explain all the assumptions you made to find this estimate. Explain why these assumptions are reasonable and whether your estimate would most likely be too large or too small.

Exercise 4. With the conditions of Exercise 3, can you put an upper bound on the size of the spill? That is, can you state with certainty the largest amount of oil that could have been spilled?

OIL SPILL INCIDENTS IN THE OCEAN

Some oil spills have occurred after a tanker is filled and at sea. In fact, some of these spills have been quite large—more than two million gallons of oil. According to one source, "such spills have occurred worldwide at a rate of three to five per year since ... 1967" [17, p. 3]. This same source states that "it is certain that oil spills will occur again" [17, p. 1].

The spill that alerted the world of a need for a policy and procedure for cleaning up oil spills was the Torrey Canyon incident. In 1967, the tanker *Torrey Canyon* spilled about 38 million gallons of oil off the shores of Great Britain and France. Damage to marine life and the shores and beaches was great. Another spill attracting a great deal of public attention was the *Exxon Valdez* spill. In 1989, this tanker hit a reef and dumped about 11 million gallons of oil into the Prince William Sound. Over 300 square miles of sea and 1100 miles of beach were contaminated. Table 1 is a chart of but a few examples of spills in the last 30 years.

We will limit our focus to spills such as those in Table 1, that is, spills at sea. We will discuss the fate of the oil once it hits the sea. There are a variety of natural processes that affect the oil on the sea—changing its make-up and determining the area contaminated. There are different clean-up techniques, some mechanical, some biological, and some chemical.

MEASURING OIL

If you check the financial section of the newspaper, you will notice current prices for a barrel of oil. Those in the oil business commonly measure oil by the barrel, where a *barrel* of oil is 42 gallons of oil. Oil is also measured in metric tons. One metric ton is about 7.35 barrels.

TABLE 1 **Examples of Spill Incidents**

Date	Ship's Name	Approximate Number of gallons spilled	Extent of coverage
1967 [13, p. 16]	Torrey Canyon	38 million	contaminated several hundred miles of shorelines
1972 [18, p. 23]	Tien Chee	2 million	contaminated 300 square miles of sea
1974 [18, p. 22]	Metula	16 million	contaminated 1000 square miles
1975 [18, p. 23]	Showa Maru	1 million	10 square mile oil slick that contaminated islands, beaches, and dock areas
1975 [18, p. 23]	Jakob Maersk	26 million	contaminated 20 miles of coastline
1975 [18, p. 23]	Epic Colocotronis	16.5 million	
1978 [13, p. 16]	Amoco Cadiz	68.4 million	
1980 [17, p. 4]	Juan A. Lavalleja	11 million	
1983 [17, p. 4]	Castillo de Bellvier	50–80 million [some oil burned]	
1983 [13, p. 16]	Pericles	14 million	
1985 [13, p. 16]	Nova	21.4 million	
1989 [15, p. 42]	Bahia Paraiso	200,000	10 square mile oil slick
1989 [15, p. 40] [19, p. 12]	Exxon Valdez	11 million	oil slick of 150 square miles contaminated over 3000 square miles of water and 1100 miles of beach
1989 [13, p. 16]	Khark-5	23.5 million	

Exercise 5. For each of the spills recorded in Table 1, determine the number of barrels of oil spilled.

To define or use models for the predicted size of an oil slick, we need to relate gallons to cubic inches, and cubic inches to cubic miles.

Exercise 6. Consider the following conversions:

$$\text{one gallon} = 231 \text{ cubic inches}$$

and

$$100,000 \text{ gallons} = 23,100,000 \text{ cubic inches}$$
$$= 13368.056 \text{ cubic yards}$$
$$= 2.452055175 \times 10^{-6} \text{ cubic miles}.$$

How many cubic miles is 1 million gallons? Eleven million gallons?

PROPERTIES OF OIL

Before we examine the fate of the oil after it is spilled, we will look at four properties of oil: density, viscosity, surface tension, and solubility. These properties have a definite impact on the fate of the oil.

The **density** of a fluid is defined as the ratio of its mass to its volume. The density of water is 1000 kg/m^3, whereas the density of crude oil ranges between 700–980 kg/m^3. Thus, oil is less dense than water, so a given volume of water weighs more than an equal volume of oil. When a certain volume of oil enters the ocean, it displaces the same volume of water. The water then buoys up or exerts an upward lifting force on the oil. According to Archimedes' principle, this force is equal to the weight of the water that was displaced. Thus, the water exerts a positive upward force on the oil and supports it; i.e., the oil floats on the water. As the oil weathers, the density of oil increases and the oil will sink below the surface.

Viscosity is the resistance of a fluid to change in its movement or shape; it is like the internal friction of a fluid. A fluid with a low viscosity flows more easily than one with a high viscosity. The viscosity of a fluid depends upon the type of fluid and the temperature. An increase in temperature lowers the viscosity of a fluid. Consider a spill of water versus a spill of the same amount of syrup. The water spreads more rapidly and covers a larger area than the syrup. The water puts up less resistance. Likewise, honey is more viscous than light corn syrup. As mentioned, temperature plays a role in viscosity. Consider a spill of hot fudge versus a spill of fudge syrup right from the refrigerator. The hot fudge spreads much more rapidly that the cold fudge syrup.

Exercise 7. Graph a plausible relation between temperature and viscosity.

Surface tension is a measure of the force of attraction between the surface molecules of a liquid, which creates a boundary surface between the liquid and another substance. When one splashes milk while pouring it, droplets of milk form. This is because of the boundary surface between the milk and the air. As another example, consider the process of blowing a bubble using a simple wire ring dipped in soap solution. One carefully removes the ring and observes a film of soap solution stretched across the ring. Surface tension holds the solution in place. If one blows too hard, when trying to form a bubble, the film is punctured and the solution is pulled by surface tension to the sides of the ring.

Exercise 8. Graph a plausible relation between temperature and surface tension. Graph a plausible relation between temperature and spreading.

Solubility is the characteristic of a substance, called the solute, which describes how it dissolves in another substance. Solubility is measured in grams of solute dissolved per liter of solution at a specified temperature. Kool-aid powder is very soluble in water. Sugar is very soluble in hot tea, but not in cold tea. That is why most people prefer an artificial sweetener for their iced tea. Oil is also not very soluble in water; thus the oil floats on the surface, with only a few droplets falling into the water.

An important force that affects the fate of a spill is **interfacial tension**. This term means the "friction force" between two substances, such as water and air, water and oil, or oil and air. Consider silk and a smooth piece of plastic. There is little interfacial tension between these two bodies, whereas there is a lot of interfacial tension between a rubber gripper and a bottle top. We rely on this tension to get the caps off stubborn jars!

Oil Hits the Ocean

When oil is spilled in the ocean, it does not sink, but rises to or remains on the surface of the sea. An important question is how large is the area of the resultant oil slick and how far does it travel. **Spreading** is the process of the oil forming into slick(s) on the sea. When we speak of the spread of a spill, we mean the oil slick resulting from the spill. We use the term **drift** to describe the movement of the slick across the surface of the sea. A slick may travel many miles, contaminating a large area of water and land.

Spreading

It seems reasonable to assume that the larger the volume of the oil spilled the larger the size of the slick. Is the oil slick size solely dependent on the volume of the oil spilled?

Exercise 9. For the spills for which the size of the slick is given in Table 1, make a graph of the number of gallons spilled (in millions) versus the size of the oil slick. Is there a linear relationship between the number of gallons spilled and the size of the slick? Explain your answer by offering a reason for your conclusion.

You have seen that the size of an oil slick is not a linear function of the volume spilled. We need to make additional assumptions in order to predict slick size. The assumptions we will make are that the sea is calm and that the spill is the size and nature of the *Valdez* Spill. Under these assumptions, some data is gathered and the spread of the spill, that is, the slick size, is modeled by the following bar graph.

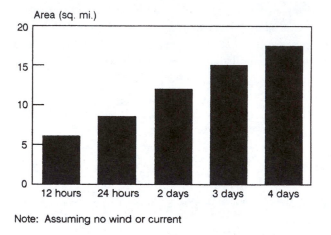

Note: Assuming no wind or current

Figure 1. Spread of an *Exxon Valdez*-sized oil spill (from [17, p. 11]).

As a first step toward getting an algebraic model (i.e., an equation) of spread, complete the following exercise.

Exercise 10. Plot the information given in Figure 1 with the x-axis representing time in half-days and the y-axis representing the number of square miles covered. Explain mathematically why the points do not fit a line. What conclusion would you draw about the spread of oil, if the data fit a line?

Now let us try to get a function that fits the data.

Exercise 11. What kind of function do you think these points suggest? Try some monomials with integer or fractional powers.

Goodman and Mac Neill [7, p. 144–145] describe an equation that J.A. Fay developed to model spread under the assumptions of calm seas and skies. Fay further assumed that the main forces causing spread are gravity, inertia, friction, and surface tension. Fay broke down the spreading of oil into three phases, naming these phases according to the main forces acting at that time. The phases are called the gravity-inertia phase, the gravity-viscous phase, and the surface-tension-viscous phase. The rate of spreading in the first two phases is dependent on the volume of the oil spill, because the force of gravity is a major force. In the last phase, when the major forces are surface tension and viscosity (properties of the oil), the volume of the oil has little effect on the spreading rate [7, p. 144–145]. The length of time in each phase depends on the size of the spill. Larger spills will spend

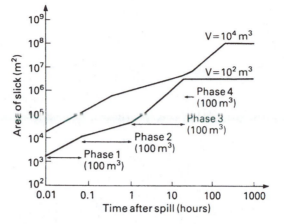

Figure 2. Influence of spill volume on spreading rate. The two volumes are $V_1 = 1000$ cubic meters and $V_2 = 10,000$ cubic meters (from [5, p. 29]).

more time in the first two phases than smaller spills [5, p. 28–29]. Figure 2 is a piecewise linear graph that shows this influence of spill volume on spreading rate and length of time in each phase. For each phase, a linear approximation is used to model the spread during that phase. The slope of the line segment is the spreading rate for that phase. Notice that the slopes vary according to the phase and volume of oil spilled. The last phase shown in the graphs represents the time after spreading ceased. Observe the segments in Phase 4 are horizontal line segments, thus having a slope of 0, consistent with the stop of spread. This piecewise linear model clearly marks the distinct phases as well as differences in spreading rates.

Exercise 12. Using the line graph in Figure 2, give the length of time the two different volumes of oil spent in each phase.

Fay's algebraic model for the area of spreading or slick size (as described by Goodman and MacNeill [7, p. 145]) is not defined piecewise but as one formula:

$$A = \omega t^{3/2}$$

where A is the area of the slick, ω is a coefficient that reflects interfacial tension and gravity, and t is time after the spill [7, p. 145]. The coefficient ω is best determined experimentally as it depends upon the type of oil.

Exercise 13. For a spill with $\omega = 6$ and t measured in days, graph the curve $A = 6t^{3/2}$. How many square miles will be covered in 2 days?

Exercise 14. Let $\omega = 8$ and measure t in days. Graph the curve $A = 8t^{3/2}$. On this same graph, plot the points obtained from the Figure 1 data. How does the curve compare to the plotted points? When do they show the greatest agreement? Try other values for ω and repeat the above directions. Which value for ω do you think gives the best fit and why?

Because different forces dominate the spread of the oil at different times, the coefficient ω is in reality a function of time [7, p. 155].

Exercise 15. Using the data given in Figure 1, find good values for ω for days 2 and 3.

When an oil slick is not treated the rate of spread decreases with time. Goodman and MacNeill [7, p. 155] describe a model that takes into account this fact. The equation is

$$A = \delta t^{2/3}$$

where A is again slick area and t is time after the spill.

Exercise 16. For $\delta = 8$, we have $A = 8t^{2/3}$. Graph this function and compare it to the data given in Figure 1. Which function ($A = 8t^{2/3}$ or $A = 8t^{3/2}$) gives the better fit? Try other values for δ and compare the results with the data given in Figure 2. Which value do you think gives the best fit? Explain.

According to the models we have shown, the area of the spread would increase without bound. In fact, untreated oil does not spread indefinitely and Fay [7, p. 155] proposes the slick reaches a maximum area of

$$A = 10^5(V)^{3/4},$$

where $V = $ the volume of the oil spilled measured in meters cubed. For example, suppose 1.5 cubic meters of oil are spilled. Then the expected slick size is about 135,540 square meters.

Exercise 17. Compute the expected slick area for a spill of 1 million gallons. Of 38 million gallons. Of 68 million gallons. You need the conversion factor that 100,000 gallons = 3780 cubic meters. Also note that 1 square mile = 2.59 square kilometers.

Exercise 18. Compute the expected slick area for the *Bahia Paraiso* spill.

Exercise 19. Compute the expected slick area for the *Valdez* spill.

In fact, the *Valdez* oil slick covered 150 square miles. The contrast in the predicted slick size and the actual slick size show that a better model would include the influence of sea and weather states. Moreover, there are many processes influencing the fate of the oil. In addition to spreading, these include drift, evaporation, photo oxidation, emulsification, dissolution, dispersion, biodegradation, and sedimentation. These processes are collectively called *weathering processes*. Each process will be defined and discussed later in the paper. There is qualitative understanding and some quantitative understanding of these other processes that affect the oil. Thus, it is very difficult to generate a model which takes all these factors into account.

As indicated above, there are many processes and forces acting on an oil spill. The following graph shows the time span and relative magnitude of the processes acting on spilled oil. This graph gives you an idea of how the forces interact with each other.

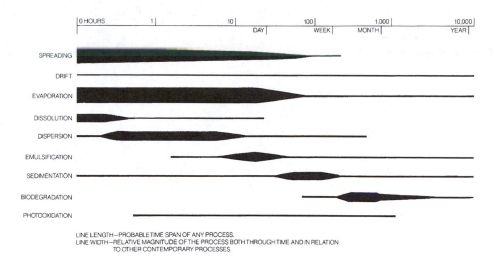

LINE LENGTH—PROBABLE TIME SPAN OF ANY PROCESS.
LINE WIDTH—RELATIVE MAGNITUDE OF THE PROCESS BOTH THROUGH TIME AND IN RELATION
TO OTHER CONTEMPORARY PROCESSES.

Figure 3. A Graph of the time span and relative magnitudes of processes acting on spilled oil (from [6, p. 6]).

Looking at the first line of Figure 3, we see the spreading is the biggest force acting on the slick at the start. The effect of spreading is greatest during the first 7 days or so, and by 10 days spreading seems to have little effect. The rate of spreading decreases as time goes on. This fact is consistent with the claim that the area covered by a slick has a limiting value.

Exercise 20. For each process listed in the graph, indicate when it is at its strongest and when it is not a major force. This should tell us what forces we can and cannot ignore at a certain time period in modeling the fate of the oil in the sea.

Drift

The second line of the graph in Figure 3 relates to drift. *Drift* is the movement of the center of a mass of oil on the surface of the ocean. Drift is caused by winds, currents, waves, and tides. For very large spills, drift is independent of the spill volume, spreading, and weathering. Drift is a process that is always acting on the oil slick with the same intensity, according to Figure 3 (and as you have observed in the previous Exercise).

To understand drift, let us look at wind and currents. Both wind and currents have a speed (a measure of size) and a direction. For example, suppose the oil spill occurred in the Northern Hemisphere where winds flow from the southwest. Southwest is the direction of the wind. The speed of wind is measured in knots where one knot is 1 nautical mile per hour. A nautical mile is 6076 feet. For example, a complete description of a wind is "the wind comes from the southwest at a speed of 15 knots."

Quantities that have both a direction and size are called vectors and can be represented by arrows. We will thus call $\vec{V}_W$ and $\vec{V}_C$ the wind and current vectors, respectively.

The contribution of the wind to the flow of the oil is given by the formula $\xi \times \vec{V}_W$. The coefficient ξ is found by computing

$$\xi = \sqrt{\frac{\rho_a}{\rho_{\text{oil}}}},$$

where $\rho_a = $ the density of air (1.3 kg/m^3) and $\rho_{\text{oil}} = $ the density of oil, which ranges between 700 to 980 kg/m^3 [5, p. 32–33].

Exercise 21. Verify that these figures give a value for ξ of 3–4%.

For example, if the wind is traveling at 15 knots, then the drift velocity due to the wind is between .45 and .60 knots. In the following example, we use the average of the values: .525.

Suppose the current moves at an angle of 45 degrees to the right of the wind. For our example the direction of the current is to the east. The speed of currents ranges between 0 and 3 knots. Assume we have an easterly current at 1.5 knots. The graph below shows how we can picture these forces of wind and current. The arrow indicates the direction and the length indicates the speed. The arrow for wind drift is $\vec{V}_w$ and the arrow for current is $\vec{V}_C$.

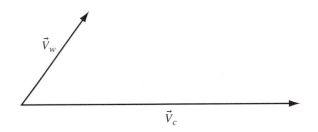

Figure 4. Graph of the wind drift and current vectors.

To find the direction and speed of the drift of the oil slick, form a parallelogram. The long diagonal vector gives the speed and direction of the slick. Call the long diagonal $\vec{V}_S$, the slick vector. This vector represents the slick velocity.

These calculations can be done algebraically. Let the origin be the beginning point for the two arrows. Then the coordinates of $\vec{V}_C$ are clearly $(1.5, 0)$. The coordinates of $\vec{V}_W$ are a bit trickier to find and require the use of some trigonometry. In general, to find the coordinates of an arrow, we carry out the following steps. Let L be the length of the arrow. Let θ be the angle made by the vector and the positive side of the x-axis. Then the coordinates are $(L \cos q, L \sin q)$. For $\vec{V}_W$ we have $L = .525$ and $\theta = 45$ degrees. So the

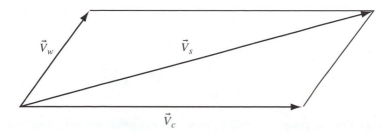

Figure 5. Graph of the slick vector.

coordinates of the tip of the arrow are

$$\left(\frac{.525\sqrt{2}}{2}, \frac{.525\sqrt{2}}{2}\right)\cdots.$$

Now to find the coordinates of the arrow representing the oil slick's movement, add the coordinates of $\vec{V}_W$ and $\vec{V}_C$. We get

$$\vec{V}_s = \left(1.5 + \frac{.525\sqrt{2}}{2}, \frac{.525\sqrt{2}}{2}\right),$$

which is approximately (1.87, .37). After one hour, a particle will cover a distance of

$$\sqrt{\left(1.5 + \frac{.525\sqrt{2}}{2}\right)^2 + \left(\frac{.525\sqrt{2}}{2}\right)^2} = 1.91 \text{ nautical miles.}$$

Exercise 22. For our example above, how far does a particle travel in 5 hours?

Exercise 23. Suppose the wind has a speed of 6 knots per hour and makes an angle of 60 degrees with the current vector. What is the velocity of a slick particle?

Evaporation

Evaporation is a weathering process that is a major clean-up force in the first day or two after the spill. The graph in Figure 3 shows that the rate of evaporation remains constant about 18 hours and then decreases until the rate stabilizes after 5 or so days.

In the first two days, some of the more toxic components of the oil evaporate. Certain refined oil products such as gasoline, light kerosene, and aviation fuels may evaporate completely. However, for most crude oils, evaporation can account for less than half of the oil spill's volume. The graph below shows how oil type influences the percentage of oil that evaporates.

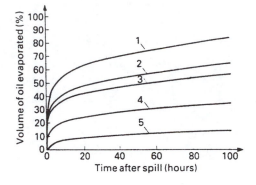

Figure 6. Relationship between time after spillage and percentage of oil evaporated for different crude oils. The results are for 5 different crude oils with wind speed of 8 knots and air temperature of 10 degrees Celsius (50° F) (from [5, p. 37]).

Exercise 24. Explain how the graph in Figure 6 shows the pattern of evaporation rate exhibited in Figure 3.

Aside from oil type, other factors influencing evaporation are the volume of the spill, wind velocity, and temperature. Figure 7 shows variations in the amount of oil evaporated caused for different volumes of spills.

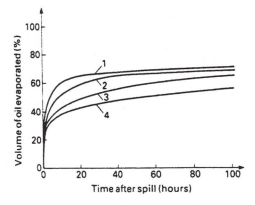

Figure 7. Relationship between the time after spillage and percentage of oil evaporated for different spill volumes. The different volumes were (1) 10 m^3, (2) 100 m^3, (3) 1000 m^3, (4) 10,000 m^3. The wind speed was 8 knots and the air temperature was 10 degrees Celsius (50° F) (from [5, p. 35]).

Exercise 25. Using the graph in Figure 7, describe the relation between the volume of a spill and the amount of oil evaporated.

Strong winds result in more evaporation. The variation in the percentage of oil evaporated at different wind speeds is shown in the graph below.

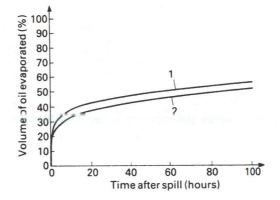

Figure 8. Relationship between the time after spillage and percentage of oil evaporated for different wind speeds. The different speeds were (1) 8 knots and (2) 2 knots. The volume of oil spilled was 104 m^3. The wind speed was 8 knots and the air temperature was 10 degrees Celsius (50° F) (from [5, p. 36]).

Exercise 26. Evaporation is enhanced by large spreadability. Given this statement, find an algebraic or graphical model of the amount of oil evaporated in relation to viscosity. Do the same for the amount of oil evaporated in relation to temperature.

The difficulty in getting a quantitative model for evaporation as a function of time is that evaporation rates depend upon many factors, all of which may vary with time [12, p. 276].

Emulsification

Emulsification is the process of water droplets being mixed into the oil slick. The percentage of water in the emulsion may increase up to a maximum of 80%. The emulsion is called *chocolate mousse*, as it is often brown, though it can also be shades of red or orange. The volume of the mousse is four times that of the starting oil. The mousse has a density approaching that of the surrounding water.

It is believed that other weathering processes such as evaporation and dissolution precede the process of emulsification [12, p. 279]. Temperature also impacts the formation of mousse. Lower temperatures are more favorable to stable emulsions. Once emulsification has begun, it has a negative effect on the other weathering processes. Evaporation and dissolution are less likely to occur. The rate of spreading also decreases.

Mousse forms regardless of sea conditions, but is formed more rapidly in choppy seas. The motion of the strong waves causes the boundary between the layer of oil and the water to deform and become unstable so that mixing occurs. Crude oils, which begin with a higher viscosity that refined oils, become even more viscous because of weathering. This factor and the content of wax in crude oil make this type of oil more likely to form stable emulsions.

Exercise 27. At the time of the *Valdez* oil spill, the seas were choppy and the winds were gusty. Use this fact and the above information to explain why the *Valdez* oil slick did not grow to its predicted size.

Dispersion

Dispersion is the process whereby the oil is "broken" into droplets which then can sink into the ocean. Thus this process is most important in terms of the break-up and disappearance of the slick.

Exercise 28. Explain how dispersion and emulsification are "opposite" process.

This process can happen naturally, caused by the presence of certain compounds in the oil. Waves can also cause dispersion, if the wind speeds are greater than 10 knots. When the force of the waves is greater than the interfacial tension, the oil slick breaks into droplets. The smaller droplets become part of the water column and the larger ones float back to the surface to rejoin the slick. Chemical surfactants can also be used to cause dispersion.

Certain properties of the slick affect the rate of dispersion. The rate of dispersion is greater for thinner slicks as wave breaking occurs more frequently. (Thicker slicks tend to dampen the waves.) If the difference between the density of the oil and the density of the water is small, then dispersion is more likely to occur. Thus, more dense oils will have greater dispersion rates than those with lower densities.

As the viscosity of the oil increases over time and emulsification takes place, the dispersion process slows down and eventually stops.

Exercise 29. Show how the above statement is consistent with Figure 3.

It is hard to develop a mathematical model for dispersion as the necessary fluid mechanics is not yet well understood [12, p. 279].

Dissolution

Dissolution is the process of the oil slick dissolving into the ocean waters. This process can have serious effects on the marine life. However, dissolution plays little role in the changing the size of the oil slick, as very little (less than a few percent) oil dissolves [12, p. 277]. This fact, plus the fact that much of the oil that is dissolved comes from dispersion imply that a separate model for dissolution is not critical [12, p. 277].

METHODS OF CLEANUP

In the previous sections we have discussed how oil spills on water travel without human intervention. In this section we will cover some methods of human intervention in the removal of oil from water after a spill. Such methods are constantly being developed, researched, and improved. Thus we will only take a cursory look at a few methods. In the United States, clean up of oil spills is required by law. This clean up is monitored and assisted by the Coast Guard. In fact, docks that receive and ship oil products have strict requirements on availability of emergency response equipment and procedures. These requirements are also enforced by the Coast Guard.

Concern over water pollution from oil spills became so intense in the mid 1970s that in 1977 the Senate Committee on Appropriations through its Subcommittee on Transportation and Related Agencies requested a study on the Coast Guard's response to oil spills. The report was delivered by the Comptroller General in May of 1978. Among many other items, the report contained information on the number of oil pollution incidents occurring in American waters during the calendar years 1975 and 1976. In 1975 there were 10,141 such incidents resulting in the spill of 14.3 million gallons of oil while in 1976 there were 10,660 incidents resulting in the spill of 23.1 million gallons. It was also reported that one spill of 7.3 million gallons was the major cause for the increase between the two years [4, p. 1]. The report detailed the source of the 1976 spills in a table similar to the one shown in Figure 9.

Exercise 30. What was the average spill size in 1975? In 1976?

Exercise 31. What percent of the 1976 total resulted from the one large spill?

Exercise 32. If the large spill were removed from the 1976 total what would have been the percent change from 1975 to 1976?

Exercise 33. Based on the 1976 table, what can you conclude about the source of the single large spill?

Source of Spill	% of incidents	% of gallons spilled
Vessels	29.1	45.9
Land Vehicles	3.9	2.0
Non-transportation related facilities	26.9	29.5
Pipeline	5.9	18.9
Marine facilities	4.8	1.4
Land facilities	1.6	1.5
Miscellaneous or unknown	27.9	0.8

Figure 9. Oil Pollution Incidents in 1976 (from [4, p. 1]).

While the body of the Comptroller General's report contained much information regarding the effectiveness of the Coast Guard in cleanup activities, Appendix I beginning with the heading "State of the Art of Oils Spill Cleanup" was most interesting. This appendix went on to say, "Although each spill is unique due to the numerous variables such as type of product spilled, geographic location of spill, weather conditions, availability of equipment, etc., there are 4 basic steps that should be taken in responding to any spill.

1. The continued flow of oil should be shut off or stopped, if possible.
2. The spill must be constrained to prevent further spreading and to lessen environmental damage.
3. The oil should be removed from the water.
4. The oil should be properly disposed of." [4, p. 52]

    The report then went on to elaborate on the "state of the art" of these items—much of which is still in use today. In the following sections we will study methods of containment and removal (items 2 and 3 of the report.)

Containment Procedures

Containment refers to the process of keeping an oil slick from spreading to locations where it will cause more environmental damage or be harder to clean up. In some cases, however, methods of containment are used to direct the oil slick into a specific location where again it will cause less environmental damage or be easier to remove from the water.

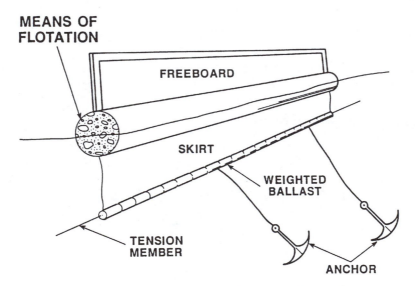

Figure 10. Basic Components of a Boom (from [11, p. 78]).

    The most common method of containment is with a mechanical barrier called a boom. While there are many different types and sizes of booms they generally share the following basic components:

1. A buoyant material or an air-fillable compartment.
2. A skirt called the float, which extends below the float to prevent oil from escaping beneath the boom. A ballast (weights) is usually attached to the skirt to hold it down in the water.

3. A section that extends above the float to keep oil from escaping over the top of the boom, usually called the freeboard.

4. A tension member to give the boom strength to withstand the forces of currents, waves, and wind.

Booms used to contain oil in the open sea are designed to withstand waves of up to 12 feet. These booms are towed in various configurations to herd the oil into smaller areas so that it can be recovered by devices to be discussed later. The configuration used is sometimes dictated by the method of recovery and is always important to successful containment. Figure 11 shows three configurations that have been found to be effective.

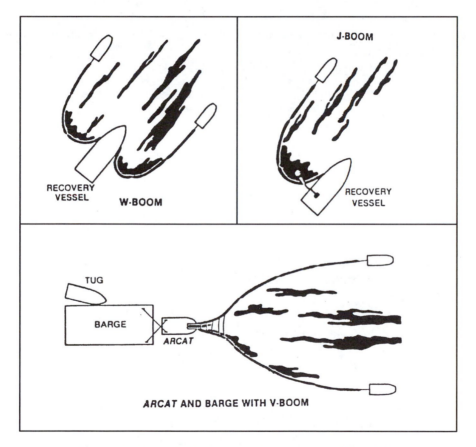

Figure 11. Three effective boom configurations (from [11, p. 82]).

The J-boom is attached to a towing vessel and the recovery vessel. Both move slowly through the water concentrating the oil near the recovery vessel, which can recover to concentrated slick more efficiently. The W-boom functions in a similar manner except two towing vessels are used and the recovery vessel is not attached to the boom. The V-boom is used with a special recovery vessel, called the ARCAT, which skims the concentrated slick off the water.

Clearly, tow speed is a critical variable in the effectiveness of any of these configurations. This variable might best be understood by considering the effect of currents on inland streams. The only difference is which is moving—the boom or the water. If the tow speed or water current is too fast, oil will escape beneath the boom's skirt. While the actual velocity at which this occurs depends on the skirt depth, and properties of the oil such as its viscosity and specific gravity, it is generally accepted that booms are ineffective in currents that exceed 1 knot, approximately 1.15 mph [11, p. 80]. In cases where the current exceeds 1 knot the boom may be deployed at an acute angle to the current, thus diverting the oil to areas of lower current or to some holding basin for eventual cleanup. *Oil Spill Response Guide* gives the following table of maximum boom angle deployment in relation to current. The angles are measured from the direction of the current so that an angle of 90° would indicate that the boom is perpendicular to the current.

Current (knots)	Current (ft/sec)	Maximum boom angle (degrees)
1.1	1.9	50
1.4	2.4	45
1.6	2.7	40
1.7	2.9	35
2.0	3.4	30

Figure 12. Current vs. boom deployment angle (from [11, p. 80]).

Exercise 34. Convert a current of 2.2 feet per second to knots, then compare your result to the first row of Figure 12. Do you think the error in this table was in recording 1.1 knots or 2.2 ft/sec? Why?

Exercise 35. A 50-foot-wide stream flows from north to south with a current of 1.6 knots. A boom, spanning the river, is to be deployed according to the recommendations in Figure 12. Find the minimum length of boom needed.

Exercise 36. If the flow in Exercise 35 is to be diverted to a point on the east side of the stream, find the location on the west bank to anchor the boom.

Extensive testing has been done on the effectiveness of booms as related to current and deployment configuration. In 1978 one such study, using the B. F. Goodrich Seaboom, was conducted by the United States Environmental Protection Agency [2, p. 4]. The tests were conducted in the USEPA's Oil and Hazardous Materials Simulated Environmental Test Tank. Nine different boom configurations were tested, ranging from "32° single diversionary" to three types of "parabolic nozzle configuration." The booms were attached to a "Bridge" situated over the test tank that towed the boom, thus simulating a current equal to the tow speed. The parabolic nozzle configurations were similar to the V-Boom configuration shown in Figure 11, while single diversionary configurations simulated deployment as discussed in Exercise 35 and 36. Two different angles of deployment were tested, 32° and 20° [2, p. 8–29]. Figure 13 summarizes the results of these tests.

Boom angle deployment (degrees)	Tow speed (m/s)	Oil escaping beneath boom %
32	0.70	1
32	0.81	5
32	0.91	10
32	1.01	50
32	1.26	70
20	0.56	0
20	0.81	0
20	1.01	7
20	1.26	10
20	1.52	30
20	1.67	50

Figure 13. Single diversionary boom test results (from [2, p. 26–27]).

Exercise 37. Compare the effectiveness of the Boom angles at a tow speed of 1.26 m/s.

Exercise 38. Plot the percent of oil that escaped beneath the boom (y) against the tow speed for:

(a) 32 degree deployment
(b) 20 degree deployment

Exercise 39. What relationship do you see between tow speed and oil escaping beneath the boom?

Exercise 40. Can you predict a tow speed for which 100% of the oil would escape for each of the deployment angles?

Exercise 41. Estimate the maximum tow speed that would allow none of the oil to escape beneath the boom for each deployment angle.

Exercise 42. Discuss situations where it might be advantageous to use faster tow speeds even though more oil would escape beneath the boom.

Mechanical Cleanup Procedures

When oil is spilled on water it floats, at least for a limited period of time. Thus the most common way of cleanup is to skim it off the water surface. This is usually done with some sort of a pump but the most efficient method is dependent on the conditions. Some types of crude oil turn into a solid when spilled in cold water and can be picked up in "frozen" sheets like ice. Most types remain a liquid and can be pumped off with a specialized skimmer pump, much like debris is skimmed off the top of a swimming pool. The most efficient type of skimmer depends on the thickness of the spill and the condition of the water. Water with broken ice seriously handicaps most skimmer activities. In these cases a device called an oil mop may be effective.

An oil mop looks much like a large diameter frayed rope. It can be deployed from the recovery vessel in a continuous loop. The mop is then dragged across the water collecting oil. The loop is then put through a wringer removing the oil into a holding tank. It is then looped back out onto the water so that the process is continuously repeated. These mops can also operate effectively in marshes, shallow water, and water containing debris.

The adjustable Weir skimmer works in much the same manner that one skims the grease off the top of a pot of chicken broth. It is shown in Figure 14. The lower lip, or Weir, floats below the oil-water interface. The depth at which the weir floats can be adjusted according to the thickness of the oil slick. When it is propelled through the water, oil, with some water, flows inside and is pumped to holding tanks. The devise works best in calm water at low tow speeds. Using it in choppy waters would be similar to skimming your chicken broth while it is boiling. It has a higher recovery efficiency when propelled at low speeds and for thick slicks. Higher speeds would increase the recovery rate, rate at which the oil is recovered, at the sacrifice of recovery efficiency or the percent of oil in the recovered liquid. The recovery efficiency for a slick that is at least 25 mm thick will average 50%. This means that the liquid recovered averages 50% oil and 50% water. However, if the thickness ranges from 1 to 8 mm the recovery efficiency averages just 10% (10% oil and 90% water) [11, p. 103]. So, just as skimming your chicken broth, the proportion of grease in the removed liquid is greater when the skimming action is slow and the layer of grease is thick.

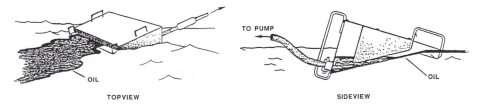

Figure 14. Weir Skimmer, (Slurp Type) (from [11, p. 104]).

Exercise 43. The data in Figure 15 represent the percent of oil in the recovered liquid (recovery efficiency) of a particular floating weir at different slick thicknesses and tow speeds.

	Towing speed 0.4 m/s				
thickness	5mm	10mm	15mm	20mm	25mm
recovery efficiency	9	30	51	71	80
	Towing speed 0.6 m/s				
recovery efficiency	5	26	47	58	74
	Towing speed 0.8 m/s				
recovery efficiency	2	19	32	51	65

Figure 15. Tests of the Efficiency of a Weir Skimmer.

(a) Plot, on one coordinate axis, the recovery efficiency (%) against slick thickness for each of the three tow speeds. This should give you three curves on the same axis.

(b) Use the plots from part (a) to predict recovery efficiency for a tow speed of one meter per second for each slick thickness.

The disc skimmer consists of a series of discs that rotate through the water. "As these discs contact the oil it adheres to them and is subsequently removed by stationary blades that continuously wipe the disc" [11, p. 115]. The removed oil is then pumped into a holding tank. Disc skimmers tend to have a low recovery rate but their recovery efficiency tends to be quite high. However, they are susceptible to damage from floating solids. The Morris MI-30 disc skimmer (Figure 16) has 30 discs made of polyvinyl chloride and a 95% recovery efficiency [11, p. 115].

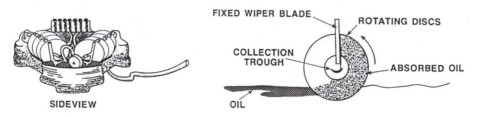

Figure 16. Morris MI-30 Disc Skimmer (from [11, p. 116]).

Exercise 44. A disc skimmer with 95% recovery efficiency can recover 110 gallons of liquid in an hour. For the weir skimmer with recovery efficiencies as reported in Figure 15, the amount of liquid recovered depends on the tow speed. At a speed of 0.4 m/s the rate is 300 gallons per hour and it increases by 50 gallons per hour for each additional 0.1 m/s tow speed.

(a) For each combination of slick thickness and tow speed reported in Figure 15 calculate the amount of oil that would be recovered in an hour.

(b) For which combinations of slick thickness and weir skimmer tow speed will it have a higher hourly oil recovery rate than the disc skimmer?

Chemical and Biological Cleanup Procedures

It is known that many species of sea-dwelling microbes have the ability to consume oil, but this method of cleanup is generally too slow to deal with large spills. In cleaning up the beaches of Prince William Sound after the *Exxon Valdez* spill of 1989 the EPA attempted to speed up the microbes' consumption process. The beaches were sprayed with a "bacteria fertilizer" in order to increase their natural ability to consume oil [15, p. 51]. Researchers are active in the development of bacteria that can consume oil and other toxic wastes. In the 1980s Ananda M. Chakrabarty and colleagues at University of Illinois Medical Center in Chicago reported success in the development of bacteria that can consume some specific toxic wastes.

Another, more researched, method of cleanup involves the use of chemical dispersants. A chemical dispersant works much like the detergents that are used for laundry or dishes. The chemical is sprayed on the oil causing it to become soluble in water or in some cases settle to the bottom. However, it is possible that the chemicals may be more hazardous to marine life than the oil itself. Thus great care must be taken in the decision to use such a method. If the chemical is used on an inland body of water, where mixing and dilution are quite slow, it may be more likely to cause a problem than if used in the open sea. On the other hand, the contaminated area may be easier to control for certain inland situations.

Ongoing research regarding the use of chemicals to control oil spills abounds. Active in this research is the American Society for Testing and Materials (ASTM), which has held several symposia for researchers to present their results. At the 1994 ASTM symposium Donald Mackay presented "Effectiveness of Chemical Dispersants Under Breaking Wave Conditions," in which he was critical of the reluctance of officials to use chemical dispersants. In the introduction of this paper Mackay begins, "Despite considerable research, review and discussion of the effectiveness and toxicity of chemical dispersants, the existence of contingency plans, and experience in field trials and actual spill incidents, there has remained a reluctance, at least in North America, to apply chemical dispersants to oil spills. An example was the *Exxon Valdez* incident in March 1989 in which only a limited amount of the available dispersant quantity was applied. In this study an assessment is made of the likely dispersion which would have resulted had the available dispersant been applied following that incident. Of particular relevance is the storm which developed on the afternoon of Sunday March 26 which created turbulent, breaking wave conditions in Prince William Sound, which would, it is believed, have enhanced dispersion" [10, p. 311].

In this four-part study Mackay, under conditions similar to those for the *Exxon Valdez* spill, developed a correlation equation for predicting the percentage of dispersed oil from the dispersant to oil ratio (DOR) and used it to demonstrate how effective dispersant application could have been in the *Valdez* spill. His equation is $Y = 100 \exp(-k/D)$, where y is the percent dispersed, D is the DOR, and k is a constant, with units of dispersant to oil ratio, which depends on the energy level. For his experimental conditions, turbulent waters, a value of 0.0015 was suggested for k [10, p. 325].

Exercise 45. Mackay's study was designed to show that relatively small amounts of dispersant would have been needed to disperse a large percentage of the oil in the *Exxon Valdez* spill. Use his equation to estimate the percent dispersed for a 1:100 DOR (i.e., when $D = 0.01$).

Exercise 46. Plot the graph of Mackay's equation for a range of DORs from 1:300 to 1:25. Use your graph to suggest a best DOR level for these conditions. At your suggested DOR, how much dispersant would be needed for a spill of 60,000 gallons?

Exercise 47. In Mackay's the value of k depends on among other things the type of oil and the wave conditions. In order to test its sensitivity to different values of this variable, plot the graph for $k = 0.001$ and 0.002. Use these graphs to discuss the models sensitivity to varying assumptions on the value of this parameter.

REFERENCES

1. Willard Bascom, *Waves and Beaches: the Dynamics of the Ocean Surface*, Doubleday & Company, Inc., Garden City, New York, 1964.

2. Michael K. Breslin, *Boom Configuration Tests for Calm-Water, Medium-Current Oil Spill Diversion*, EPA-600/2-78-186, August 1978. (Prepared by Mason & Hanger–Silas Mason Co. for the U.S. Environmental Protection Agency.)

3. Steven H. Cohen and William T. Lindenmuth, *Design, Fabrication and Testing of the Air-Jet Oil Boom*, EPA-600/7-79-143, June 1979. (Prepared by Hydromatics Incorporated for the U.S. Environmental Protection Agency.)

4. Comptroller General of the United States, *Coast Guard Response to Oil Spills: Trying to Do Too Much With Too Little*, United States General Accounting Office, Washington, D.C., 1978.

5.] J.W. Doerffer, *Oil Spill Response in the Marine Environment*, Pergamon Press, New York, 1992.

6. Exxon Corporation, *Fate and Effects of Oil in the Sea*, Exxon Corporation, 1985.

7. Ron H. Goodman and Margaret R. MacNeill, "The Use of Remote Sensing in the Determination of Dispersant Effectiveness," in *Oil Spill Chemical Dispersants*, Tom Allen, Ed., ASTM, Philadelphia, 1992, 143–160.

8. Douglas J. Graham, Robert W. Urban, Michael K. Breslin, and Michael G. Johnson, *OHMSETT Evaluation Tests: Three Oil Skimmers and a Water Jet Herder*, EPA-600/7-80-020, February 1980. (Prepared by PA Engineering and Mason & Hanger–Silas Mason Co. for U.S. Environmental Protection Agency.)

9. George M. Hidy, *The Waves: The Nature of Sea Motion*, Van Nostrand Reinhold Co., New York, 1971.

10. Donald Mackay, "Effectiveness of Chemical Dispersants Under Breaking Wave Conditions," in *The Use of Chemicals in Oil Spill Response*, ASTM STP 1252, Peter Lane, Ed., American Society for Testing and Materials, Philadelphia, 1995, 310–340.

11. Robert J. Meyers and Associations and Research Planning Institute, Inc., *Oil Spill Response Guide*, Noyes Data Corporation, Park Ridge, New Jersey, U.S.A., 1989.

12. National Research Council, *Oil in the Sea: Inputs, Fates, and Effects*, National Academy Press, Washington, D.C., 1985.

13. National Research Council, Committee on Tank Vessel Design, *Tanker Spills: Prevention by Design*, National Academy Press, Washington, D.C., 1991.

14. National Research Council, Committee on Effectiveness of Oil Spill Dispersants, *Using Oil Spill Dispersants on the Sea*, National Academy Press, Washington, D.C., 1989.

15. *Science Year 1991: The World Book Annual Science Supplement*, World Book, Inc., Chicago, 1991.

16. Gary F. Smith, *Performance Testing of Selected Sorbent Booms*, EPA-600/7-78-219, November 1978. (Prepared by Mason and Hanger–Silas Mason Co. for the U.S. Environmental Protection Agency.)

17. U.S. Congress, Office of Technology Assessment, *Coping With an Oiled Sea: An Analysis of Oil Spill Response Technologies*, OTA-BP-0-63, U.S. Government Printing Office, Washington, D.C., 1990.

18. U.S. Environmental Protection Agency, Oil and Special Materials Control Division, WH-548, *Oil Spills and Spills of Hazardous Substances*, Washington, D.C., 1977.

19. Russell G. Wright, *Oil Spill!* (Student Edition), Addison-Wesley Publishing Co., Menlo Park, California, 1995.

Geometry Measures Tank Capacity and Avoids Oil Spills

Yves Nievergelt
Eastern Washington University

INTRODUCTION: MATHEMATICS AVOIDS CHEMICAL SPILLS

Used in the *design* of storage tanks and their operating procedures, mathematics helps effectively in preventing subsequent environmental damage, for instance, material waste or chemical spills.

As an example of an environmentally effective mathematical design, the present note demonstrates the calculation of charts and graduated dip sticks to measure the amount of liquid chemical contained at the bottom of a storage tank, and, by subtraction, the volume still available to fill up the tank without overflow.

The problem just described arose in a telephone inquiry from American Transport's office in Spokane, Washington, on 3 January 1991. Based in Portland, Oregon, American Transport delivers fuel oils by trucks to underground tanks at customers' plants, for example, at gas stations. Upon delivery, the truck driver first lowers a vertical graduated ruler—like a car's dip stick—into the tank to measure the amount of fuel in the tank, as in Figure 1, which determines the remaining space available for delivery. Such precaution will enable the driver to stop the truck's pump before overfilling the underground tank, without spilling oil. To this purpose, however, the driver needs a means to convert the reading of depth on the vertical ruler into an estimate of the volume of fuel in the tank, for example, with a conversion chart. Unfortunately, such charts cannot be readily produced through experiments at each new customer's tank. Is there then a formula to generate such conversion charts, or a formula programmable into a truck on-board computer?

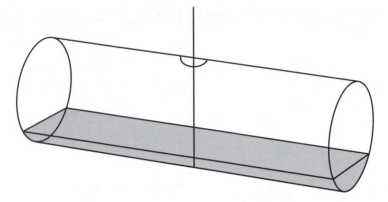

Figure 1. Calculate the volume of the fuel in terms of the depth of the fluid.

Observe that the problem neither involves nor asks for any number; instead, it asks for a literal formula. Also, notice that the problem does not come with a picture, partly because it came via telephone, and partly because everyone in the oil business knows what a cylindrical tank looks like.

Other examples and exercises will reveal that the same mathematics also applies to the management of municipal water tanks and to the monitoring of nuclear-waste storage tanks.

Indexed by pairs of indices (k, ℓ), the following subsections may be read in any non-decreasing order of the sum $k + \ell$, for instance, subsections $(1, 1)$, $(1, 2)$, $(2, 1)$, $(2, 2)$, or $(1, 1)$, $(2, 1)$, $(1, 2)$, $(2, 2)$.

1 MATHEMATICS MEASURES THE VOLUME OF FUEL IN TANKS

The present section demonstrates how to calculate the volume of fuel in a storage tank in terms of the fuel's depth and the tank's dimensions. To simulate real life, some of the material appears here as exercises but does not include solutions. After the first subsection, which requires only introductory geometry and algebra, the second subsection solves American Transport's problems with introductory calculus.

1.1 Tanks Involving Introductory Geometry without Calculus

The following exercises simulate the situation described in the introduction, but with tanks of shapes that require introductory spatial geometry, without calculus.

Exercise 1. Consider a rectangular (parallelepipedic) tank with horizontal length ℓ, horizontal breadth b and vertical depth d. Suppose that the tank contains some liquid chemical, which fills only a part of the tank, from the bottom up to a height h above the bottom of the tank. Derive a formula for the remaining volume still available for additional chemical in the tank.

Exercise 2. Consider a tank in the shape of an inverted vertical pyramid with a square horizontal cross section, with its vertex at the bottom, horizontal breadth b at the top, and vertical depth d. Suppose that the tank contains some liquid chemical, which fills only a part of the tank, from the bottom up to a height h above the bottom of the tank. Derive a formula for the remaining volume still available for additional chemical in the tank.

The following exercises pertain to tanks of shapes that may also involve the number π.

Exercise 3. (This problem arose in the management of the water tank of the town of Union in Oregon [8]. The same problem also applies to the monitoring of nuclear-waste storage tanks [14].) Consider a tank in the shape of a vertical right circular cylinder, with diameter (horizontal breadth) b and height (vertical depth) d. Suppose that the tank contains some liquid chemical, which fills only a part of the tank, from the bottom up to a height h above the bottom of the tank. Derive a formula for the volume of chemical in the tank.

Exercise 4. Consider a tank in the shape of a vertical right circular cone, with its vertex at the top, horizontal breadth b at the bottom, and vertical depth d. Suppose that the tank contains some liquid chemical, which fills only a part of the tank, from the bottom up to a height h above the bottom of the tank. Derive a formula for the remaining volume still available for additional chemical in the tank.

Exercise 5. Explain why the ratio a/r^2 of the area to the squared radius is the same for all discs. In other words, explain from the definition of area why the same number π works for all discs.

Exercise 6. Taking for granted that the area a of a disc of radius r equals πr^2, explain from the definition of area why the area of an ellipse with principal semi-axes of lengths r and s equals $\pi r s$.

1.2   Tanks Involving Calculus

The present section demonstrates how to solve the initial problem posed by American Transport.

Problem

Imagine a fuel tank in the shape of a horizontal cylinder, with length ℓ and circular cross-section with diameter d. Moreover, suppose that fuel partially fills the tank, from the bottom to a height (depth) h above the bottom. Find a formula for the volume of fuel in the tank, denoted by $V(\ell, d, h)$.

Solution.   The following solution shows what skills students need to solve applied calculus problems. Firstly, a picture may help intuition, and sketching one by hand may take less time than with a computer (see Figure 2). Secondly, translating the problem from its

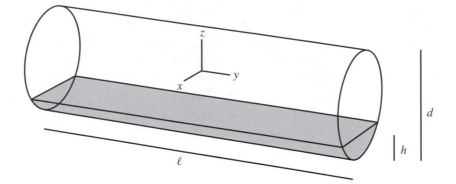

Figure 2. Calculate the volume of the fuel in terms of ℓ, d, and h.

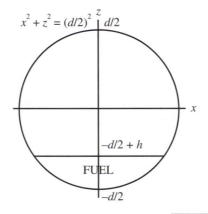

Figure 3. The fuel lies where $-d/2 \le z \le -d/2 + h$ and $-\sqrt{(d/2)^2 - z^2} \le x \le \sqrt{(d/2)^2 - z^2}$.

geometric statement into a calculus exercise cannot yet readily be done by a computer, and, consequently, must still be done mentally. To this end, identify the cross-section of the tank with the disc with diameter d and center at the origin, characterized by the in-equation $x^2 + z^2 \le (d/2)^2$. Thus, the fuel in the tank occupies the region bounded by the inequations (see Figures 2 and 3)

$$\begin{cases} -\ell/2 \le y \le \ell/2 & \text{(for a total length } \ell) \\ -d/2 \le z \le -d/2 + h & \text{(depth } z \text{ from bottom to height } h), \\ -\sqrt{(d/2)^2 - z^2} \le x \le \sqrt{(d/2)^2 - z^2} & \text{(width } x \text{ between the tank's walls).} \end{cases}$$

Thus, the fuel's volume takes the integral form

$$V(\ell, d, h) = \int_{y=-\ell/2}^{y=\ell/2} \int_{z=-d/2}^{z=-d/2+h} \int_{x=-\sqrt{(d/2)^2-z^2}}^{x=\sqrt{(d/2)^2-z^2}} dx \, dz \, dy, \tag{0}$$

Either with such a package as *Mathematica*, through the command (see [12, p. 630–636])

```
Integrate[ Integrate[ Integrate[ 1,
  { x, -Sqrt[(d/2)^2-z^2], Sqrt[(d/2)^2-z^2] } ],
  { z, -d/2, -d/2+h } ], { y, -ℓ/2, ℓ/2 } ]
```

or by hand, *routine* calculus yields

$$V(\ell, d, h) = \ell * \left\{ \left(\frac{d}{2}\right)^2 * \left(\frac{\pi}{2} - \text{Arcsin}\left(1 - \frac{2h}{d}\right)\right) + \left(h - \frac{d}{2}\right) * \sqrt{h * (d - h)} \right\}.$$

(1)

A simplification reduces the need for two arithmetic operations, and for π, which many computers do not offer: inverse to the trigonometric identity $\cos(\{\pi/2\} - \alpha) = \sin(\alpha)$, the identity

$$\frac{\pi}{2} - \text{Arcsin}(t) = \text{Arccos}(t)$$

leads to the alternate formula

$$V(\ell, d, h) = \ell * \left\{ \left(\frac{d}{2}\right)^2 * \text{Arccos}\left(1 - \frac{2h}{d}\right) + \left(h - \frac{d}{2}\right) * \sqrt{h * (d - h)} \right\}. \quad (2)$$

Because of the importance of formula (2) in practice, as already explained, and because of the unpleasant consequences that an error may cause, several verifications appear appropriate. Such verifications form the object of the following problems, which, to simulate reality, appear here without answers.

Exercise 7. (Alternate derivation.) Arrive at formula (2) in a different manner, by integrating not the areas of the vertical cross-sections, but the areas of the rectangular horizontal slices of the fuel. What integral emerges instead of integral (0)? What theorem about mulitple integrals do the result of the present problem and formula (2) illustrate?

Exercise 8. (Geometric tests.) Select a few values of h for which you can calculate the volume $V(\ell, d, h)$ geometrically, for example, $h = 0$ (empty tank), $h = d/2$ (half-empty tank), and $h = d$ (full tank). Compare the expressions of $V(\ell, d, h)$ obtained through geometry with those given by formula (2).

Exercise 9. (Numerical tests.) American Transport has requested sample values over the telephone to verify their implementation of formula (2) on their computer. Program formula (2) on a programmable calculator. Consider a real tank 395 inches long and 95 inches in diameter. Verify with your calculator that such a tank has a capacity of about 12,000 gallons. Also, how many gallons of fuel should American Transport's computer indicate for a height of one inch at the bottom of the same tank?

Exercise 10. (Trigonometric variants.) Some versions of BASIC offer neither Arccos nor Arcsin, but only Arctan, abbreviated as ATN (see [3]). To accommodate such compilers, rewrite formula (2) with an expression involving Arctan but not Arccos, by means of trigonometric identities similar to those used to derive formula (2) from formula (1).

Besides the preceding routine problems, for instance, for a term project, students may attempt the following extensions of problem 1, which aim at providing students with additional experience in solving real problems by means of elementary calculus.

The project described here consists of four problems, which address further inquiries from American Transport at increasing levels of difficulty.

On 17 August 1991, American Transport wrote to request a formula similar to (2) but for "fiberglass tanks that are shaped like a giant Tylenol capsule," which forms the object of the following problem.

Exercise 11. (Hemispherical ends.) Establish a formula, $T(\ell, d, h)$, for the volume of fuel with depth h at the bottom of a horizontal cylinder with length ℓ, circular cross section of diameter d, and with *hemispherical caps* at both ends. In other words, the cylindrical tank considered here does not end with a vertical planar disc, but with a half of a sphere at each end. Thus, the *total* length of the tank is the sum $\ell + d$ of the length of the cylindrical part, ℓ, and the diameter d, because the two hemispheres each add a radius, $d/2$, to the total length. Then test your formula with the dimensions of a real tank of this type, with a length of $\ell = 329$ inches and a diameter of $d = 96$ inches: your formula should yield a volume $T(425, 96, 96)$ of *about* 12,000 gallons for the full tank.

On 13 February 1991, American Transport had called again, requesting assistance in calculating the volume of fuel in tanks with "oval" cross sections.

Exercise 12. (Elliptical section.) Establish a formula, $W(\ell, a, b, h)$, for the volume of fuel with depth h at the bottom of a horizontal cylinder with length ℓ and elliptical cross section with horizontal diameter a and vertical diameter b. The ends of the tank are vertical flat elliptical pannels. Test your result with the dimensions of a real tank, in inches: $\ell = 172$, $a = 90, b = 63$, and $h = 90$ (full tank).

Exercise 13. (Ellipsoidal ends.) In the preceding problem, also take into account the ends of the tank, which are not vertical flat walls, but ellipsoidal caps with a bulge of three inches at the center. Thus, the total length of the tank along its axis is $3 + 172 + 3 = 178$ inches at the center, but still 172 inches along the perimeter.

Exercise 14. (Approximation of "oval" section.) For a more accurate approximation of reality, take into account the fact that the cross section is not exactly elliptical. For the real tank in problem 7, the perimeter of the cross section is an "oval," symmetric with respect to the origin, and passing through the points in the first quadrant with coordinates $(45, 0)$, $(40, 15.75)$, $(22.5, 28)$, and $(0, 31.5)$.

For a greater challenge, students may try Strang's version of problem 1: with a tank neither horizontal nor vertical, but tilted [11]. Documentation for such a challenging calculus problem appears in Gray's note [4], supported by the firm Martin Marietta Energy Systems, where the design of a gauge for a storage tank in a chemical processing plant requires the calculation of the volume of water inside a tilted cylindrical pipe, either through calculus by hand or with the software Maple.

Alternatively, students may read Balk's account of independent calculations, first by the Venerable Cooper, and later by Kepler in 1615, to measure the *full* capacity of Linz's wine barrels by dipping a stick from the middle of the side diagonally to the opposite internal "corner" of the barrel.

2   COMPUTER AND CALCULATOR INACCURACY MIGHT CAUSE SPILLS

Though formulae exist to calculate the volume of fluid in a tank, as demonstrated by the preceding exercises, digital computers and calculators cannot evaluate such formulae exactly:

> My suspicion is that the vast majority of code for geometric algorithms is faulty for these reasons.   —Joseph O'Rourke [10, p. 91].

Hence arises the question whether computational inaccuracies may lead to erroneous conclusions and then spills. As a typical example, the present section shows how to estimate the accuracy of the computation of the area and volume of a tank, which also demonstrates the usefulness of inequalities.

2.1   Algebraic Estimates of Calculator Accuracy

The built-in inaccuracies of digital computers and calculators arises because for comparisons and calculations people often prefer numbers in digital representations. Yet some numbers involve infinitely many digits, which require an infinite time to calculate and display. Therefore, digital computers and calculators by design process only a fixed finite number of digits, which requires only a finite time but introduces inaccuracies in the results.

Example 1   A decimal ten-digit calculator computing $1/(\frac{1}{7})$ produces not 7 but $6.999\,999\,998$:

$7.000\,000\,000$	enter 7.
$\boxed{1/x}$	press the "reciprocal" key.
$0.142\,857\,142\,9$	displayed approximation of $\frac{1}{7}$.

$\boxed{1/x}$ press the "reciprocal" key.

6.999 999 998   displayed approximation of $1/\left(\frac{1}{7}\right)$.

Different calculators produce different inaccuracies. For instance, a decimal twelve-digit calculator computing $1/(1/7)$ produces $7.000\,000\,000\,01$.

While an error in the tenth or twelfth digit may seem inconsequential, such errors compound after several operations in sequences of computations.

Example 2 A decimal ten-digit calculator computing $1/\{[1/(\frac{1}{7})] - 6.999\,999\,999\}$ produces not the exact result $1{,}000{,}000{,}000$ but $-1{,}000{,}000{,}000$ instead. In contrast, a decimal twelve-digit calculator computing $1/\{[1/(\frac{1}{7})] - 6.999\,999\,999\}$ produces $2{,}000{,}000{,}000$ instead. *Observe that an initial error in the tenth digit caused an error in the first digit of the final result.*

To reduce the effects of computational inaccuracies, a definition of accuracy will prove useful.

Definition 1. The **absolute error** of an approximation $\tilde{w}$ of a nonzero number w is the absolute value of their difference:

$$|\tilde{w} - w|.$$

The **relative error** of an approximation $\tilde{w}$ of a nonzero number w is the ratio

$$\frac{|\tilde{w} - w|}{|w|}.$$

Example 3 A decimal ten-digit calculator computing $1/(\frac{1}{7})$ produces a result $(6.999\,999\,998)$ with an absolute error

$$|\tilde{w} - w| = |6.999\,999\,998 - 7| = 2 * 10^{-9}$$

and a relative error smaller than 10^{-9}:

$$\frac{|\tilde{w} - w|}{|w|} = \frac{|6.999\,999\,998 - 7|}{|7|} = \frac{2 * 10^{-9}}{7} < 10^{-9}.$$

Example 4 A decimal ten-digit calculator computing $1/\{[1/(\frac{1}{7})] - 6.999\,999\,999\}$ produces a result $(-1{,}000{,}000{,}000)$ with an absolute error $2{,}000{,}000{,}000$,

$$|\tilde{w} - w| = |(-1{,}000{,}000{,}000) - 1{,}000{,}000{,}000| = 2{,}000{,}000{,}000,$$

and a relative error 2, which means that the error is twice as large as the exact result (1,000,000,000):

$$\frac{|\tilde{w} - w|}{|w|} = \frac{|(-1,000,000,000) - 1,000,000,000|}{|1,000,000,000|} = \frac{2,000,000,000}{1,000,000,000} = 2.$$

Most pocket calculators use base ten, and most digital computers use base two, but some digital machines use other bases, for instance, three, eight, or sixteen [6]. For each arithmetic operation or each evaluation of an elementary function, digital computers and calculators conforming to IEEE standards [7] produce an absolute error of at most one half in the last displayed digit, or a relative error of at most one half in the penultimate displayed digit [5, p. 179–180]. This means that if the computer or calculator displays d digits in base b, then the displayed result $\tilde{w}$ approximates the exact result w with an absolute error

$$|\tilde{w} - w| \le \left(\frac{1}{2}\right) * b^{-d},$$

and a relative error

$$\frac{|\tilde{w} - w|}{|w|} \le \frac{1}{2} * b^{-d+1} = \frac{b}{2} * b^{-d}.$$

Example 5 A decimal ten-digit calculator has a built-in function that produces an aproximation $\tilde{\pi} = 3.141\,592\,654$ of the number π with an absolute error of at most $(\frac{1}{2}) * 10^{-10}$, so that [5, p. 184]

$$|\tilde{\pi} - \pi| = |3.141\,592\,654 - \pi| \le \tfrac{1}{2} * 10^{-10},$$

which means that

$$3.141\,592\,653\,5 \le \pi \le 3.141\,592\,654\,5.$$

In particular, a digital computer or calculator replaces the arithmetic operations $+$, $-$, $*$, $/$ by approximations $\oplus$, $\ominus$, $\circledast$, $\oslash$.

Example 6 A decimal ten-digit calculator computes nonzero multiplications with relative errors smaller than $(\frac{10}{2}) * 10^{-10} = 5 * 10^{-10}$. Thus, with nonzero numbers u and v already stored in the calculator, the calculator approximates their product $w := u * v$ by a number $\tilde{w} := u \circledast v$ with a relative error

$$\frac{|(u \circledast v) - (u * v)|}{|u * v|} = \frac{|\tilde{w} - w|}{|w|} \le 5 * 10^{-10}.$$

Example 7 Consider the problem of computing the area a of a disc with positive squared radius r^2, for instance, the area of the circular base of a cylindrical storage tank:

$$a = \pi * r^2.$$

In such a computation, a digital calculator introduces two approximations: $\tilde{\pi}$ for π, and $\circledast$ for $*$. Because of the absolute error $|\tilde{\pi} - \pi| \leq (\frac{1}{2}) * 10^{-10}$, it follows that $\tilde{\pi} r^2$ differs from πr^2 by

$$|\tilde{\pi} * r^2 - \pi * r^2| = |\tilde{\pi} - \pi| * r^2 \leq \left(\frac{1}{2}\right) * 10^{-10} * r^2.$$

Because of the relative error in the multiplication, $\tilde{\pi} \circledast r^2$ differs from $\tilde{\pi} * r^2$ by

$$\frac{|(\tilde{\pi} \circledast r^2) - (\tilde{\pi} * r^2)|}{|\tilde{\pi} * r^2|} \leq 5 * 10^{-10}.$$

Consequently, the calculator approximates the area $a = \pi * r^2$ by $\tilde{a} := \tilde{\pi} \circledast r^2$ with a relative error

$$
\begin{aligned}
\frac{|\tilde{a} - a|}{|a|} &= \frac{|(\tilde{\pi} \circledast r^2) - (\pi * r^2)|}{|\pi * r^2|} \\[2mm]
&= \frac{|(\tilde{\pi} \circledast r^2) - (\tilde{\pi} * r^2) + (\tilde{\pi} * r^2) - (\pi * r^2)|}{|\pi * r^2|} \\[2mm]
&\leq \frac{|(\tilde{\pi} \circledast r^2) - (\tilde{\pi} * r^2)| + |(\tilde{\pi} * r^2) - (\pi * r^2)|}{|\pi * r^2|} \\[2mm]
&= \frac{|(\tilde{\pi} \circledast r^2) - (\tilde{\pi} * r^2)|}{|\pi * r^2|} + \frac{|(\tilde{\pi} * r^2) - (\pi * r^2)|}{|\pi * r^2|} \\[2mm]
&\leq \frac{|(\tilde{\pi} \circledast r^2) - (\tilde{\pi} * r^2)|}{|\pi * r^2|} + \frac{\frac{1}{2} * 10^{-10} * r^2}{|\pi * r^2|} \\[2mm]
&= \frac{|(\tilde{\pi} \circledast r^2) - (\tilde{\pi} * r^2)|}{|\pi * r^2|} + \frac{1}{2\pi} * 10^{-10} \\[2mm]
&= \frac{|(\tilde{\pi} \circledast r^2) - (\tilde{\pi} * r^2)|}{|\tilde{\pi} * r^2|} * \frac{|\tilde{\pi} * r^2|}{|\pi * r^2|} + \frac{1}{2\pi} * 10^{-10} \\[2mm]
&\leq 5 * 10^{-10} * \frac{|\tilde{\pi} * r^2|}{|\pi * r^2|} + \frac{1}{2\pi} * 10^{-10} \\[2mm]
&= 5 * 10^{-10} * \frac{|\tilde{\pi} * r^2 - \pi * r^2 + \pi * r^2|}{|\pi * r^2|} + \frac{1}{2\pi} * 10^{-10} \\[2mm]
&\leq 5 * 10^{-10} * \left(\frac{|\tilde{\pi} * r^2 - \pi * r^2|}{|\pi * r^2|} + \frac{|\pi * r^2|}{|\pi * r^2|}\right) + \frac{1}{2\pi} * 10^{-10} \\[2mm]
&\leq 5 * 10^{-10} * \left(\frac{1}{2\pi} * 10^{-10} + 1\right) + \frac{1}{2\pi} * 10^{-10} \\[2mm]
&\leq 5 * 10^{-10} * \left(\frac{1}{2\pi} * 10^{-10} + 1 + \frac{1}{10\pi}\right)
\end{aligned}
$$

$$< 5 * 10^{-10} * \left(1 + \frac{1}{25}\right)$$
$$= 5.2 * 10^{-10} < 10^{-9}.$$

Thus the calculator computes the area with an error of at most one in the eighth displayed digit. Hence, to avoid overestimating the area, it suffices to round *down*, or *chop*, the calculator's result by one unit in the eighth digit. Similarly, to avoid underestimating the area, it suffices to round *up* the calculator's result by one unit in the eighth digit.

Remark 1. Digital computers and calculators may also produce errors caused by phenomena called *underflow* and *overflow*, but such errors do not occur with the type of data considered here, typically integers with at most four digits. *Underflow* occurs if the magnitude of a result is not zero but smaller than the smallest positive number representable in the machine, typically 10^{-99}, which then rounds down to zero. *Overflow* occurs if the magnitude of a result is larger than the largest positive number representable in the machine, typically 10^{99}, which then produces an error warning in the "floating-point" arithmetic of most digital computers and calculators. However, the "wrap-around" integer arithmetic generated by some C compilers does not warn of overflows but allows the result to "wrap around" by a few bits into a meaningless result, with the opposite sign but without warning [10, p. 157].

Exercise 1. Determine the relative error in the computation of the volume of a parallelepiped with length ℓ, depth d, and height h.

Exercise 2. Determine the relative error in the computation of the volume of a sphere with radius r.

Exercise 3. Determine the relative error in the computation of the volume of a right circular cylindrical tank with length ℓ and diameter d.

Exercise 4. Without the need for any calculus, the number π has been known for millenia [13] with an accuracy sufficient for the present purpose. Suppose that one knows that π lies between two known positive numbers p and q, so that $p < \pi < q$. Also suppose that one wants to avoid overfilling storage tanks, and, therefore, that one prefers to underfill such tanks slightly rather than overfilling them at all. In the various calculations of volumes of vertical cylindrical or conical storage tanks, must one then use the lower estimate p or the higher estimate q instead of the unknown exact value of π?

2.2 Calculus Estimates of Calculator Accuracy

The present subsection indicates how to estimate the accuracy of the formula for American Transport:

$$V(\ell, d, h) = \ell * \left\{ \left(\frac{d}{2}\right)^2 * \mathrm{Arccos}\left(1 - \frac{2h}{d}\right) + \left(h - \frac{d}{2}\right) * \sqrt{h * (d - h)} \right\}. \quad (2)$$

Because the manufacturer's data for d and ℓ and the driver's data for h involve only a few, typically at most four, digits, computers and calculators carry sufficiently many digits to produce the exact values of

$$\left(\frac{d}{2}\right)^2, \quad \left(1 - \frac{2h}{d}\right), \quad \left(h - \frac{d}{2}\right), \quad h * (d - h).$$

However, the values of the square root and of the inverse trigonometric functions need not be rational numbers, which means that computers and calculators may need to round their results in their last displayed digits. For displays with d digits in base b, machines conforming to IEEE standards yield an absolute accuracy such that for every real $u \geq 0$ and every real c with $-1 \leq c \leq 1$,

$$|\widetilde{\sqrt{u}} - \sqrt{u}| \leq \left(\frac{1}{2}\right) * b^{-d},$$

$$|\widetilde{\mathrm{Arccos}}(c) - \mathrm{Arccos}(c)| \leq \left(\frac{1}{2}\right) * b^{-d},$$

and a relative accuracy such that for every real $u > 0$ and every real c with $-1 \leq c < 1$,

$$\frac{|\widetilde{\sqrt{u}} - \sqrt{u}|}{|\widetilde{\sqrt{u}}|} \leq \frac{b}{2} * b^{-d},$$

$$\frac{|\widetilde{\mathrm{Arccos}}(c) - \mathrm{Arccos}(c)|}{|\widetilde{\mathrm{Arccos}}(c)|} \leq \frac{b}{2} * b^{-d}.$$

In particular, the results become relatively inaccurate only if they have small magnitudes. Consequently, digital machines evaluating formula (2) do not produce large errors: simply rounding the final result down to the nearest gallon should avoid any spill.

Exercise 5. Verify that if all the dimensions d, ℓ, and h are integers for which $0 < \ell < 10^3$, $0 < d < 10^2$, and $0 \leq h < 10^2$, with inches as units, and if a decimal twelve-digit calculator conforms to the error bounds just listed, then the machine computes $V(\ell, d, h)$ by formula (2) with an error smaller than one gallon, so that rounding down the remaining volume by one gallon avoids a spill.

Exercise 6. Suppose that a *binary* computer evaluates $V(\ell, d, h)$ by formula (2), with data as in the foregoing exercise. How many binary digits must the computer carry for a final error smaller than one gallon?

3 CONCLUSION

The case study presented here, thanks to American Transport, may contribute to a livelier calculus. Yet the case study also confirms that the skills developed in calculus texts—be they lean and lively or mean and deadly—correspond to the skills required by real applications: reading and understanding problems stated in prose, drawing figures, translating prose into mathematics, setting up multiple integrals, using algebraic or trigonometric identities and inequalities, and so forth. Furthermore, the case study shows that while calculus problems may seem dull or abstract without documentation by case studies, their applications may nevertheless be quite slick.

4 ACKNOWLEDGEMENT

This work was supported in part by the National Science Foundation's grant DUE-9455061. Classification numbers appear below to facilitate library searches.

REFERENCES

1. M. B. Balk, "The Secret of the Venerable Cooper," *Quantum*, May 1990, p. 36–39.
2. John A. Belward, "Tank Calibration," *UMAP Journal*, Vol. 15, No. 1 (Spring 1994), p. 29–42.
3. James S. Coan, *Advanced Basic: Applications and Problems*, Hayden, Rochelle Park, NJ, 1977. ISBN 0-8104-5856-X. QA76.73.B3C6. 76–7435.
4. L. J. Gray, "Storage Tank Design," *SIAM Review*, Vol. 33, No. 2 (June 1991), p. 271–274.
5. Hewlett-Packard, *HP-15C Advanced Functions Handbook*, part number 00015-90011 Rev. B, Hewlett-Packard Company, Portable Computer Division, Corvallis, OR, 1984.
6. Nicholas J. Higham, *Accuracy and Stability of Numerical Algorithms,* Society for Industrial and Applied Mathematics, Philadelphia, PA, 1996. ISBN 0-89871-355-2. QA297.H53 1996. 95-39903. 519.4'0285'51–dc 20.
7. *IEEE Standard for Binary Floating-Point Arithmetic, ANSI/IEEE Standard 754-1985*. Institute of Electrical and Electronic Engineers, New York, NY, 1985.
8. Yves Nievergelt, "Practitioner's Commentary: The Outstanding Water Tank Papers," *UMAP Journal*, Vol. 12, No. 3 (Fall 1991), p. 239–241.
9. Yves Nievergelt, "Calculus Measures Tank Capacity and Avoids Oil Spills," *College Mathematics Journal*, Vol. 25, No. 2 (March 1994), p. 132–135.
10. Joseph O'Rourke, *Computational Geometry in C*, Cambridge University Press, Cambridge, UK, 1993. ISBN 0-521-44034-3.
11. Gilbert Strang, *Calculus,* Wellesley-Cambridge Press, Wellesley, MA, 1991. ISBN 0-13-032946-0. QA303.S8839 1991. 90-49977. 515.20.
12. Stephen Wolfram, *Mathematica: A System for Doing Mathematics by Computer*, 2nd ed., Addison-Wesley, Redwood City, CA, 1991. ISBN 0-201-51502-4. QA76.95.W65 1991. 91-46832. 510'.285'53–dc20.
13. Bartel Leenert van der Waerden. *Geometry and Algebra in Ancient Civilizations*. Berlin: Springer-Verlag, 1983. ISBN 0-387-12159-5. QA151.W34 1983. 83-501. 512'.009.
14. Paul Whitney (Batelle Northwest), invited address at 9:30 a.m. on Friday 16 June 1995 at the meeting of the Pacific Northwest Section of the Mathematical Association of America at Whitman College in Walla Walla, WA.

Lead Poisoning in Humans

Robert S. Cole
The Evergreen State College, WA
Robert M. Tardiff
Salisbury University, MD

INTRODUCTION

This chapter addresses the issue of lead poisoning in humans, currently considered one of the most prevalent of childhood diseases. In fact K. W. James Rochow recently wrote: "While it (lead) is the most prevalent childhood disease in the U.S., childhood lead poisoning is an environmental disease that is completely preventable" [6, 1995, p. 90]. Rochow footnotes this statement with statistics that 10% of U.S. children under age six are lead poisoned (defined to be more than 10 μg/dl—micrograms of Pb per deciliter of blood—see Appendix 1), and among low income black children the rate is 30% [9, 1992].

One step in addressing this problem is to create a model of how lead flows through the human body. This will give some insight into how various changes in the environmental exposure to lead will affect human health. The model we will be examining is a compartment model that traces the exchange of lead between various organs of the body and the environment. It was developed by Naomi Harley and Theodore Kneip [2, 1988, pp. 27–36], and serves as the basis for a model used by the Environmental Protection Agency for estimation of body lead burdens in young children, given lead concentrations in various sources of childhood lead intakes [5, 1993, 10, 1989, 11, 1990].

This model is one of several developed in the last twenty years to trace the effects of lead poisoning in humans. Different models may be suitable for different purposes. The type of compartment model used here has multiple uses in toxicology in tracing flows of contaminants, as well as in a wide variety of biogeochemical processes in environmental studies and in numerous other settings.

229

In this chapter we illustrate the ease of converting conceptual compartment models into computerized form with the use of the software *Stella*. Visual representation of relationships among parts of complex systems is an extremely valuable first step in creating a mathematical model. The *Stella* software is designed to extract symbolic, algebraic relationships, equations, and difference equations from a diagram that contains data. The diagram is intended to be an intuitive, visual representation of a model. Not surprisingly, we have found visual representations of mathematical models, rather than the more symbolic representations, to be immediately accessible to a wider variety of people.

Another tremendous advantage of *Stella* diagrams is that they explicitly allow for connections with the environment—the diagrams can be made to represent closed-loop models of substance flow. Visual modeling makes much more apparent that wastes don't just disappear, they actually go somewhere which may or may not be taken into account in the model.

CONSTRUCTION OF THE MODEL

Most substances taken into the body enter through either the lungs or the gastrointestinal tract, and then pass into the blood stream. From there they transfer to various body organs. Standard procedure has been to consider each body organ a separate compartment, and to consider inflows and outflows of the substance of interest from that compartment/organ. Generally, the substance is assumed to be well-mixed in the compartment over time periods short compared with the biological removal time. Which organs are to be considered as separate compartments, which ones can be lumped into a single compartment, and what are the appropriate inflow and outflow channels are issues that biologists and medical personnel debate. The ultimate test of the validity of a given model is its ability to replicate the important features of empirical data from measurements on the organisms in question.

The Harley and Kneip model used in this chapter was based upon chronic oral lead exposures originally administered to infant and juvenile baboons [3, 4, 1974, 1983] and later adapted to humans [2, 1988]. There are five compartments in this model with thirteen associated variables. The five compartments are called (1) Blood, (2) Bone, (3) Liver, (4) Kidney, and (5) Gut. The variables Q_i ($i = 1 \ldots 5$) are the amounts of lead per kilogram of body weight in the ith compartment at age t. The quantities λ_{ij} ($i = 1 \ldots 5$; $j = 1 \ldots 6$), which we call proportionality functions, are the proportion of the lead in the ith compartment that flows to the jth compartment per unit time, except for λ_{16}, which is proportion of lead that flows from the blood to the outside environment per unit time (see Appendix 2). Note, the λ_{ij} can depend on age. Figure 1 is a standard visual representation of such a model and is, indeed, a slight modification of the original diagram in [2, 1988, p. 31]:

The transfer of lead between organs takes place through the intermediary of blood—there is no direct exchange of material among organs except through the blood. The model assumes the rate of flow of lead at time t from one compartment to another is the amount of lead in the donor compartment times the corresponding proportionality function; for example, the rate of flow per unit time from Kidney to Blood is $\lambda_{41}Q_4$.

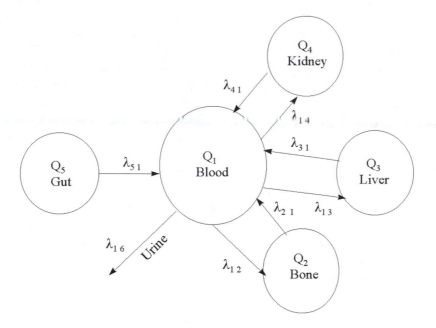

Figure 1. Standard diagram.

Since this model consists of compartments connected by flows, it is readily represented mathematically by creating a *Stella* diagram corresponding to the conceptual diagram (Figure 2).

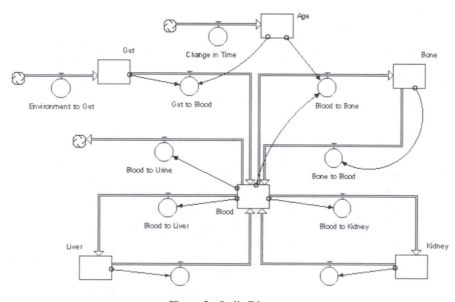

Figure 2. *Stella* Diagram.

The standard diagram and the *Stella* diagram represent the same system. In the *Stella* diagram a compartment is a rectangle and the amount of lead in a compartment at time t is stored in the variable, which also names the compartment, e.g., Blood. Lead flows from one compartment to another are indicated by a double lined arrow (pipe) with a gate valve (a gate valve is the symbol ⏣, and the rate at which lead is flowing per unit time is stored in the variable that also names the gate valve, e.g., Blood to Bone. Dependencies between variables are indicated by a single lined arrow between them with the convention that the one being pointed at depends on the other.

When we create a *Stella* diagram we are building visually the overall structure of a model and **simultaneously** the *Stella* software is creating the structure of the underlying difference equations. The *Stella* software will use the equations it generates to simulate the system's behavior. At some point the user may want to examine these equations to get insight into equilibrium states (more on this later).

The actual mathematical functions that dictate the various flows and the initial conditions are included in the rectangles and gate valves and can viewed and edited by double clicking the mouse on the rectangle or valve. To indicate that the initial amount of lead in the Liver is 0 we would enter 0 into the Liver rectangle. To indicate the rate at which lead flows from blood to kidney is proportional to the amount of lead in the blood with proportionality constant equal to 0.03, we would enter 0.03*Blood into the Blood_to_Kidney gate valve. In the model we are considering, the proportion of lead in the blood that flows to the bone per unit time is regarded as a step function (see Appendix 2). To model this in *Stella* we enter

$$(\text{if (Age} < 20*365) \text{ then } 0.34 \text{ else } 0.11)*\text{Blood} \tag{1}$$

into the Blood to Bone gate valve. This step function is an simplification of how lead flow varies with age. In Exercise 4 you will experiment with more realistic functions and observe their effects on the model's performance.

Once all the flows and initial conditions have been entered, *Stella* will have constructed a set of difference equations that describes the model analytically and will **automatically** use these equations to drive numerical simulations. The user never really needs to see these equations, but sometimes it is helpful to examine them, if for no other reason than to double check that numbers and equations have been entered correctly into the diagram. For example the code *Stella* generates for the Liver is as shown in Figure 3.

```
☐  Liver(t) = Liver(t - dt) + (Blood_to_Liver - Liver_to_Blood) * dt
   INIT Liver = 0
   INFLOWS:
       ⏣ Blood_to_Liver = .10*Blood
   OUTFLOWS:
       ⏣ Liver_to_Blood = .03*Liver
```

Figure 3. Sample *Stella* Equations.

The first line is the difference equation. The number dt is always positive (and is called t in calculus). In our simulation dt will be 1 day. INIT Liver $= 0$ means the initial amount of Lead in the liver at time $t = 0$ is 0. What follows INFLOWS: and OUTFLOWS: are the expressions that describe the rate lead is flowing into and out of the liver at time t (see Appendix 4).

RUNNING THE SIMULATION

Exercise 1. *Single Dose*   Create the *Stella* model depicted above with the data as given in Appendix 4. Assume the child is 8 years old (8×365 days) and has never been exposed to lead. What does this mean in terms of the initial conditions for each of the compartments? Next assume the child consumes a single dose of $.05\mu g$ Pb/kg (micrograms of lead per kilogram of body weight). Modify the Gut compartment accordingly. Conjecture what the lead load in Bone will look like over the next 2 years (730 days). Run a *Stella* simulation with $dt = 1$; display a graph of the lead load in the five compartments. Is the amount of lead in bone consistent with your conjecture? Explain why the lead load in bone behaves so much differently from the other compartments. What are some possible health implications? From the graph (see Appendix 3, Figure 7) it would appear that the lead load in bone peaks at about 0.04 μg Pb/kg. Increase the number of days in the simulation so that you can graphically estimate (see Appendix 3, Figure 8) how long it will take for the lead load in bone to be reduced to 0.02 μg Pb/kg.

Exercise 2. *Equilibrium*   Assume that the amount of lead entering each organ equals the amount of lead leaving each organ; i.e., the "system" is in equilibrium, and that the child is between 10 and 20 years old. Because lead is excreted from the body, to maintain equilibrium, lead must flow into the body from the environment. Assume that the rate at which lead is flowing from the environment to the gut is a. Using the symbolism in Appendix 2, verify that for the system to be in equilibrium, the following equations must hold:

Blood Equilibrium: $0 = \text{LtoB*Liver} + \text{OtoB*Bone} + \text{KtoB*Kidney} + \text{GtoB*Gut}$

$- (\text{BtoL} + \text{BtoO} + \text{BtoK} + \text{BtoU})\text{*Blood};$

Bone Equilibrium: $0 = \text{BtoO*Blood} - \text{OtoB*Bone};$

Gut Equilibrium: $0 = \text{GtoB*Gut} - a;$

Kidney Equilibrium: $0 = \text{BtoK*Blood - KtoB*Kidney};$

Liver Equilibrium: $0 = \text{BtoL*Blood} - \text{LtoB*Liver}.$

Solve these equations to show the amounts of lead in each compartment when the system is in equilibrium are:

$$\text{Gut} = \frac{a}{\text{GtoB}}; \quad \text{Blood} = \frac{a}{\text{BtoU}}; \quad \text{Liver} = \text{Blood}\frac{\text{BtoL}}{\text{LtoB}};$$

$$\text{Kidney} = \text{Blood}\frac{\text{BtoK}}{\text{KtoB}}; \quad \text{Bone} = \text{Blood}\frac{\text{BtoO}}{\text{OtoB}}.$$

Note, these solutions exhibit the **same** symmetry that the *Stella* diagram has. Blood is the "main compartment" and the liver, bone, and kidney compartments are all related to blood in the same way.

Assuming a 10- to 20-year-old child's system is in equilibrium and is receiving a constant dose of lead to the gut equal to 0.01 μg Pb/kg per day, solve for the amounts of lead in each compartment. Note how large the lead load is in Bone relative to the other compartments.

Exercise 3. *Chronic Exposure* Assume the body is receiving a constant dose of 0.05 μg Pb/kg per day. Assume the individual is 8 years old and has never been exposed to lead. Observe how the lead builds up in the system over time. Explain the "strange" behavior of the graphs at about 4,400 days (12 years) into the simulation (hint, consider the lead load in bone first.) What are possible health implications for the organs at 4,400 days (or at age 20)? Use the symbolic solution found in Exercise 2 to find the long term (equilibrium lead loads) in each compartment, assuming the person survives. If you have large enough computer capacity, run the simulation for 11,000 days, 30 years, to observe the system very slowly approaching equilibrium in bone.

Exercise 4. *Proportionality Functions* We noted in the initial discussion of the model that the use of a step function such as (1) as a proportionality function is an oversimplification. However, it is easier to see the effect of a step function has on the model than that of a more complicated function. And sometimes a simple model produces useful qualitative results.

As a first attempt at a more realistic function, describe symbolically the proportionality function for blood to bone that has value 0.34, if the person is less than 15 years old, decreases linearly from 0.34 to 0.11 as the person's age increases from 15 to 25 years, and then remains constant at 0.11 for the rest of the person's life. Next convert this expression to an equivalent expression where the unit of time is a day instead of a year. Lastly, using nested *if* statements, enter this expression into the gate valve for blood to bone, run a simulation for at least 6000 days. Comment on the effects this more sophisticated function had on the model.

While this last function is probably more realistic, note that it is still is implausible at 15 and 25 years, respectively. (Why would we expect the function to be smooth at 15 and 25 years?) A perhaps even better function would be something like that shown in Figure 4.

Finding a formula for such a function could be challenging. However, *Stella* will allow you to enter a the graph of a function and then will read the graph to determine values of the function at various ages. In fact, the graph in Figure 3 is the graph we entered into the

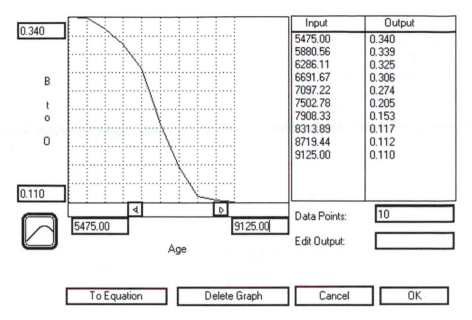

Figure 4. BtoO as a function of age measured in days.

connector (the circle) labeled BtoO on the top right side of the following *Stella* diagram (Figure 5). Experiment with different graphical functions and comment on the effects.

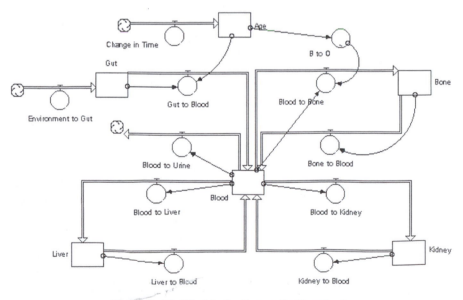

Figure 5. Modified *Stella* diagram for Exercise 4.

LEAD AND THE ENVIRONMENT

The environment plays a pivotal role in the amount of lead flowing through a human body. Two major ways lead enters the body are via the lungs and the gut.

Two sources of lead in the atmosphere are leaded gasoline and leaded paint. Leaded gasoline continues to be used in several parts of the world. It is noteworthy that the discontinuance of burning leaded gasoline in the United States resulted in an average reduction of lead in human blood by one-third to one-half [8, 1988, p. 7]. Even though the amount of lead in paint has been reduced significantly, the lead dust created by remodeling or just wear and tear in older homes with lead paint is a major source of lead. Over twelve million children in the United States under the age of seven are potentially exposed to toxic levels of lead from paint and dust [8, 1988, p. 7].

Dietary sources of lead include both water and food supply, especially wild game exposed to lead shot and root crops such as carrots and potatoes. Lead can enter the water supply by passing through older plumbing using lead solder to seal joints, or by passing through local rock formations containing lead. Lead shot consumed by birds as a digestive aid (instead of grains of sand) would induce a flow of lead through the bird's body in a manner similar to the process described for humans.

Lastly, lead can get into the soil and thus into certain crops in several ways. Sewage sludge with a lead content below 500 micrograms of Pb per gram of sludge (500 ppm) can be spread on agricultural fields (sludge with a higher lead content must be disposed of as a toxic material). Another source is fertilizer; the lead content in fertilizer (or *any* toxic material for that matter) is not regulated in commercial fertilizer products, nor in agricultural soil. Toxic wastes, including heavy metals like lead, are currently being used as ingredients in commercial fertilizers (*Seattle Times*, July 4, 1997) as a means of toxic waste disposal!

Since lead in sludge or in commercial fertilizers mixed with soils is stored in the roots of some crops, concentration of toxins in the food chain is a danger. In particular, potatoes and carrots grown on such soil become sources of dietary lead, and significant amounts of lead could be in carrot juice because of the large number of carrots necessary to make a glass of carrot juice. Lead can also lodge in the roots of grass, but not in the leaves. So animals that eat just the leaves of grass grown on lead-contaminated soil are not likely to ingest lead. However, animals that eat the leaves to the ground or some of the root may be at risk.

The lead flow from the human body to the sewage sludge to root crops back to humans is an example of feedback. This scenario would be difficult to simulate because obtaining initial conditions and flows of lead from, to, and through the environment are not easily available or readily measured. However, a *Stella* diagram for such a model would be quite useful in getting an overview and would offer some guidance as to what data should be sought.

Exercise 5. *Environment*   Construct the *Stella* diagram described in the previous paragraph. Discuss how you might expect lead to flow. Identify possible feedback loops. Make initial decisions on what sort of data you would seek.

SOME SUMMARY REMARKS

The model suggests that lead, once in the body, is very difficult to remove. Indeed a person's own bones become a source of lead poisoning even when the lead source is removed. For women, the problem may be even more serious, since accumulation of bone lead during childhood creates problems in pregnancy. Bone lead in pregnant women is passed to the fetus, and bone lead in lactating mothers is passed to infants who breastfeed [7, 1993].

The *Stella* software gives a nice visual representation of the model. However, the software can be overwhelmed by the large number of time values needed to run the simulation for a long period of time. That is where standard analytical methods (solving for equilibrium) continue to play a central role. Using a software package such as *Maple*, or numerical analysis routines, can make finding such solutions not quite so burdensome. Indeed, when examining an unknown system it is often helpful to start a simulation at or near equilibrium. Systems that behave very erratically at or near their equilibria need further analysis. The time interval dt may need to be smaller, a more sophisticated numerical method may be needed, or the system may even be chaotic.

Lastly, modeling lead flow in the environment is an ongoing research problem. Two alternative approaches described in the bibliography are Borelli and Coleman's nice exposition of the Rabinowitz model and O'Flaherty's model of lead flow in rats.

ACKNOWLEDGMENTS

We wish to thank Professors Courtney Coleman of Harvey Mudd College, Ben Fusaro of Florida State University, Stephen Gehnrich of Salisbury State University, Naomi H. Harley of New York University, Patricia Kenschaft of Montclair State University, Jude Van Buren of The Evergreen State College, and Elichia Venso of Salisbury State University for their help in the preparation of this paper.

APPENDIX 1.   UNITS

The units of lead concentration in the body are micrograms of Pb per kilogram of body mass (μg/kg). These units afford a way of measuring and comparing doses and their effects across subjects of different mass. These are ideal units to use when one is administering a known dose of lead, as was done in the original experiment on baboons [2, 1988, 4, 1983].

However, micrograms of Pb per kilogram of body mass are not the most practical units for measuring lead contamination and diagnosing potential victims of lead poisoning (can you explain why?). The most common unit for measuring lead contamination in the body is micrograms of Pb per deciliter of whole blood (μg/dl). One deciliter (a tenth of a liter) is 100 cc (cubic centimeters) since one liter is 1000 cc. A blood sample is taken (typically much less than a deciliter), the lead content of the sample is measured, and then the results are scaled up to a deciliter. Toxic effects of lead associated with various concentrations of lead in blood are shown in Figure 6 [10, 1989]. Standards for acceptable levels of lead in blood have decreased steadily over the years: 60 μg/dl was acceptable from 1960 to 1970, and has been decreased three times since then to the current 1991 standard of 10μg/dl.

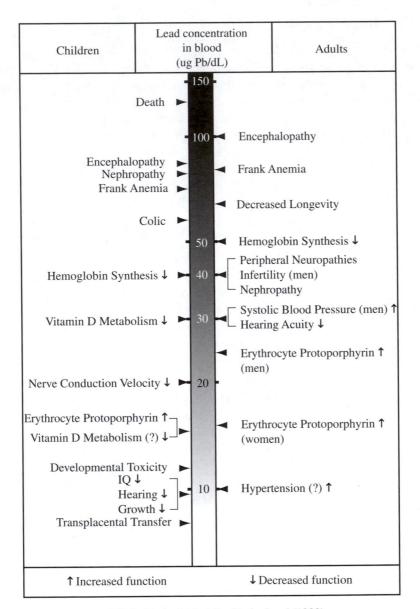

Adapted from ATSDR, Toxicological Profile for Lead (1989)

Figure 6. Effects of inorganic lead on children and adults—lowest observable adverse effect levels (from *Case Studies in Environmental Medicine*, 1992).

APPENDIX 2. PARAMETER VALUES

In the first column of the following table are standard names for the proportionality functions and in the third column, their corresponding values as reported in [4, 1983]. We also need names for these quantities that can be used in *Stella*; these names are in the second column of the table. (For example, the name for the proportionality function associated with the flow **B**lood **to K**idney is BtoK; the other names are developed similarly except for the proportionality functions associated with Blood and Bone, where we used **O** to stand for Bone.)

Standard Symbol	Derived Symbol	Value (days^{-1})
λ_{12}	BtoO	0.34, if age $\leq$ 20 years
		0.11, if age $>$ 20 years
λ_{13}	BtoL	0.10
λ_{14}	BtoK	0.03
λ_{16}	BtoU	0.08
λ_{21}	OtoB	0.00173
λ_{31}	LtoB	0.03
λ_{41}	KtoB	0.07
λ_{51}	GtoB	0.20, if age $\leq$ 10 years
		0.10, if age $>$ 10 years

APPENDIX 3. *STELLA* GRAPHS FOR EXERCISES 1, 3, AND 4

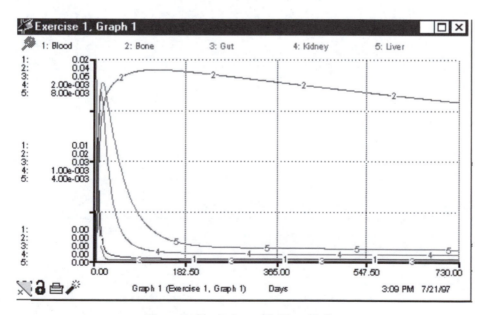

Figure 7. Single dose of 0.05 μg Pb/kg.

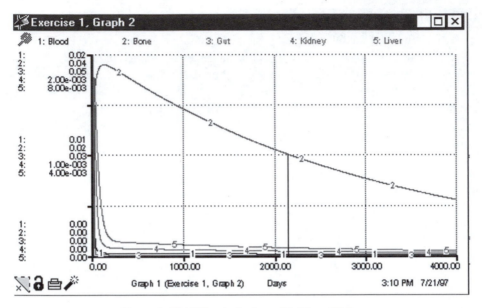

Figure 8. Half-life, single dose of 0.05 μg PB/kg.

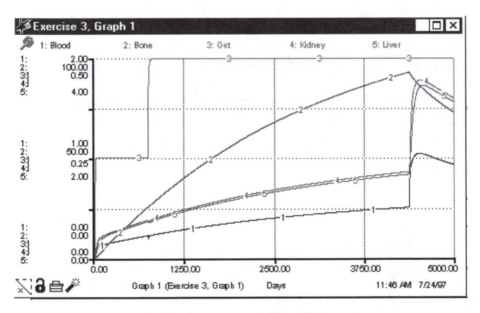

Figure 9. Chronic exposure to 0.05 μg Pb/kg per day.

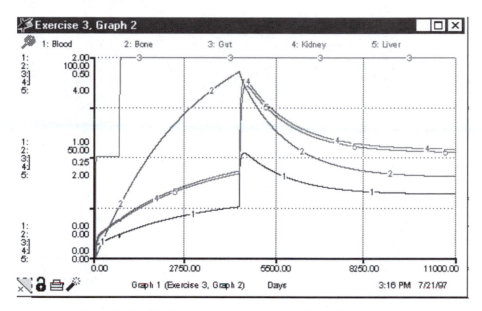

Figure 10. Equilibrium, chronic exposure to 0.05 μg PB/kg per day.

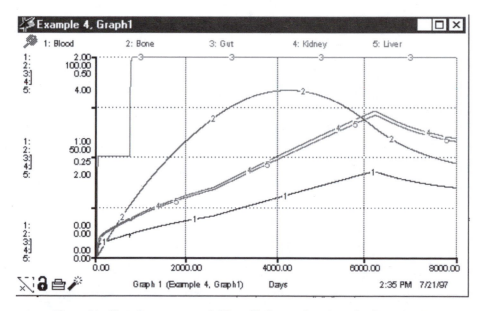

Figure 11. Chronic exposure to 0.05 μg Pb/kg per day, piecewise linear BtoO.

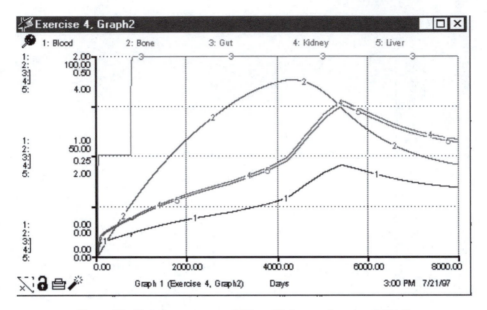

Figure 12. Chronic exposure to 0.05 μg Pb/kg per day smooth BtoO.

APPENDIX 4.   *STELLA* EQUATIONS FOR EXERCISE 1

☐ Age(t) = Age(t - dt) + (Change_in_Time) * dt
    INIT Age = 8*365
    INFLOWS:
        ♂ Change_in_Time = 1

☐ Blood(t) = Blood(t - dt) + (Gut_to_Blood + Liver_to_Blood + Kidney_to_Blood - Blood_to_Liver - Blood_to_Kidney - Blood_to_Urine - Blood_to_Bone) * dt
    INIT Blood = 0
    INFLOWS:
        ♂ Gut_to_Blood = (IF(Age<10*365) then .20 else .10)*Gut
        ♂ Liver_to_Blood = .03*Liver
        ♂ Kidney_to_Blood = 0.07*Kidney
    OUTFLOWS:
        ♂ Blood_to_Liver = .10*Blood
        ♂ Blood_to_Kidney = 0.03*Blood
        ♂ Blood_to_Urine = .08*Blood
        ♂ Blood_to_Bone = (IF(Age< 20*365) then 0.34 else 0.11)*Blood

☐ Bone(t) = Bone(t - dt) + (Blood_to_Bone) * dt
    INIT Bone = 0
    INFLOWS:

☐ Gut(t) = Gut(t - dt) + (Environment_to_Gut - Gut_to_Blood) * dt
    INIT Gut = 0.05
    INFLOWS:
        ♂ Environment_to_Gut = 0
    OUTFLOWS:
        ♂ Gut_to_Blood = (IF(Age<10*365) then .20 else .10)*Gut

☐ Kidney(t) = Kidney(t - dt) + (Blood_to_Kidney - Kidney_to_Blood) * dt
    INIT Kidney = 0
    INFLOWS:
        ♂ Blood_to_Kidney = 0.03*Blood
    OUTFLOWS:
        ♂ Kidney_to_Blood = 0.07*Kidney

☐ Liver(t) = Liver(t - dt) + (Blood_to_Liver - Liver_to_Blood) * dt
    INIT Liver = 0
    INFLOWS:
        ♂ Blood_to_Liver = .10*Blood
    OUTFLOWS:
        ♂ Liver_to_Blood = .03*Liver

Figure 13. *Stella* Equations for Example 1.

REFERENCES

1. Committee on Measuring Lead in Critical Populations, *Measuring Lead in Infants, Children and Other Sensitive Populations*, Commission on Life Sciences, National Academy of Science, National Academy Press, Washington, D.C., 1993.

2. Naomi H. Harley, "Biological Modeling for Predictive Purposes," in Methods for Biological Monitoring, *A Manual for Assessing Human Exposure to Hazardous Substances*, Theodore J. Kneip and John V. Crable, eds., American Public Health Association, Washington D.C., 1988.

3. T. J. Kneip, D.M. Goldstein, and N. Cohen, "Lead Toxicity Studies in Infant Baboons: A Toxicological Model for Childhood Lead Poisoning," in *Report to Consumer Products Safety Commission*, T. J. Kneip and E. Pfitzer, eds., Report No. CPSC-C-74-153, CPSC, Washington D.C., 1974.

4. T. J. Kneip, R. P. Mallon, and N. H. Harley, "Biokenetic Modelling for Mammallan Lead Metabolism," *Neuro Toxicology* 4 (3), 1983, 189–192.

5. Paul Mushak, "New Directions in the Toxicokinetics of Human Lead Exposure," *NeuroToxicology*, 14(2–3), 1993, 29–42.

6. K. W. Rochow, "A Pound of Prevention, an Ounce of Cure: Paradigm Shifts in Childhood Lead Poisoning Programs," in *Lead Poisoning, Exposure, Abatement, and Regulation*, Joseph . Breen and Cindy R. Stroup, eds., Lewis Publishers, CRC Press, 1995.

7. Ellen Silbergeld et al., "Lead in Bone: Storage Site, Exposure Source, and Target Organ," *NeuroToxicology*, 14(2–3), 1993, 225–236.

8. U.S. Agency for Toxic Substances and Disease Registry, "The Nature and Extent of Lead Poisoning in Children in the United States: A Report to Congress," U.S. Dept. of Health and Human Services, Atlanta, 1988.

9. U.S. Agency for Toxic Substances and Disease Registry Lead Toxicity, "Case Studies in Environmental Medicine," Herbert L. Needleman, MD, guest ed., Sarah E. Royce, MD, MPH, guest contributor, U.S. Department of Health and Human Services, September, 1992.

10. U.S. Environmental Protection Agency, "Review of the National Ambient Air Quality Standards for Lead: Exposure Analysis Methodology and Valuation," OAQPS Staff Report, Office of Air Quality Planning and Standards, Research Triangle Park, 1989.

11. U.S. Environmental Protection Agency/Science Advisory Board, "Review of the OAQPS Lead Staff Paper and the ECAO Air Quality Criteria Document Supplement," Report of the Clean Air Scientific Advisory Committee (CASAC), Washington D.C., Report No. EPA-SAB-CASAC-90-002, 1990.

BIBLIOGRAPHY

1. R.L. Borelli and C.S. Coleman, *Differential Equations: A Modeling Approach*, Prentice Hall, 1987.

2. Joseph J. Breen and Cindy R. Stroup, eds., *Lead Poisoning, Exposure, Abatement, and Regulation*, eds., Lewis Publishers, CRC Press, 1995.

3. J. Michael Davis, Robert W. Elias, and Lester W. Grant, "Current Issues in Human Lead Exposure and Regulation of Lead," *NeuroToxicology* 14 (2–3), 1993, 15–28.

4. Paul Mushak, "Biological Monitoring of Lead Exposure in Children: Overview of Selected Biokentic and Toxicological Issues," in *Lead Exposure and Child Development*, M. A. Smith, L. D. Grant, and A. I. Sors, eds., Kluwer Academic Publishers, Boston, 1989.

5. Paul Mushak, "Perspective: Defining Lead as the Premier Environmental Health Issue for Children in America: criteria and Their Quantitative Application," *Environmental Research*, 59, 1992, 281–301.

6. Ellen J. O'Flaherty, "Physiologically Based Models for Bone-Seeking Elements ii. Kinetics so Lead Disposition in Rats," *Toxicology and Applied Pharmacology*, 111, 1991, 313–331.

7. Michael B. Rabinowitz, George W. Weterhill, and Joel D. Kopple, "Kinetic Analysis of Lead Metabolism in Healthy Humans," *The Journal of Clinical Investigation*, 58, 1976, 260–270.

8. Michael B. Rabinowitz, "Toxicokinetics of Bone Lead," *Environmental Health Perspectives*, 91, 1991, 33–37.

9. Barry Richmond Barry and Steve Peterson, *Stella Technical Documentation*, High Performance Systems, Hanover, New Hampshire, 1996.

How's the Weather up There:
Predicting Weather for Coastal Mountains

Michael E. Folkoff
Donald C. Cathcart
Steven M. Hetzler
Salisbury State University, MD

1   INTRODUCTION

The focus of this chapter is forecasting weather for specific elevations on the windward and leeward sides of a mountain. This exercise is designed for upper-level high school and introductory college courses in mathematics or earth science.

2   THE MODELING PROCESS

2.1   Real World Problem

The processes in which mountains are a direct control over the weather are called *orographic processes*. Excellent examples of the effect of orographic processes can be found all over the world, including the great North-South mountain ranges of the Western United States. In the U.S., the prevailing winds are generally from the west, although there is considerable variance in this pattern due to passing storm systems embedded in the westward flow. If atmospheric conditions are favorable, the westerlies propel air over the windward western-side of the mountain and force the air to descend the leeward eastern-side. As air ascends the windward side its temperature decreases. Temperature and the moisture-holding capacity of air are positively related. As the temperature of the rising air decreases,

247

water vapor content (humidity) tends toward saturation because of the reduced moisture-holding capacity. In other words, as temperature decreases with increasing elevation, the relative humidity, which is the ratio of humidity to moisture-holding capacity, increases.

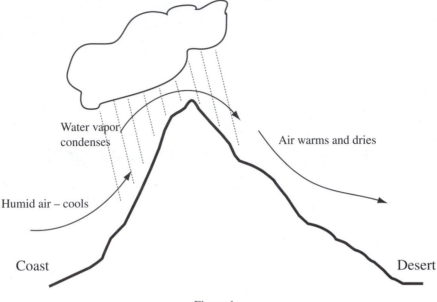

Figure 1.

Eventually, the temperature of the ascending air has cooled sufficiently so that it becomes saturated (relative humidity becomes 100%), if the air has sufficient water vapor. The elevation at which saturation is reached is called the *level of condensation* and the temperature at which air becomes saturated is called the *dew point*. Thus, at the level of condensation, the temperature and dew point are equal and remain so as long as relative humidity is 100%. Any further rise above the level of condensation keeps relative humidity at 100%, because cooling air causes the moisture holding capacity to decrease, forcing excess water vapor to condense. The condensing water forms droplets that precipitates and falls toward earth. It is the cooling and condensing of ascending air that explains the humid environment on the windward side.

Air, cresting the summit and then descending the leeward side of the mountain, is affected by the opposite forces. The descending air warms and dries (relative humidity drops below 100%) because the humidity remains fairly constant, but the moisture-holding capacity increases with increasing temperature. The decreasing relative humidity and increasing temperature due to the descending air explains the aridity of the leeward side.

Using the qualitative model described above, we develop a quantitative model that predicts the weather along a windward coast with an adjacent mountain range. Although weather conditions prevail over a wide area, weather data are only collected at discrete locations from which the model for the area is developed. Given the temperature and dew point for a location at sea level in this region, we wish to predict the temperature and dew point at various elevations on the windward and leeward sides of an adjacent mountain range. Where, if at all, does it start to rain or snow?

2.2 Problem Formulation

The discussion of physical principles suggests that the important variables are

$$e = \text{elevation above the windward base in feet}$$

$$D(e) = \text{dew point in } °\text{F at elevation } e \text{ on the windward side, and}$$

$$T(e) = \text{air temperature in } °\text{F at elevation } e \text{ on the windward side.}$$

We observe that for a given elevation, the temperature and dew point could be higher or lower on the leeward side than on the windward side. In fact in our model, leeward side temperatures and dew points will usually be higher. We therefore need separate variables to represent leeward temperatures and dew points:

$$d(e) = \text{dew point in } °\text{F at elevation } e \text{ on the leeward side and}$$

$$t(e) = \text{air temperature in } °\text{F at elevation } e \text{ on the leeward side.}$$

Relative humidity can be approximated by examining the temperature and dew point levels. We are also only concerned with whether the relative humidity is 100 percent. There are three basic environmental relationships in our model involving these variables. Many of the relations involved in atmospheric processes are not a straight line. However, it is possible to develop the model, by assuming a straight line relationship.

On the windward side, initially the air rising up the mountain side is unsaturated. As elevation increases the atmospheric pressure decreases, which cools air without a transfer of energy (*adiabatic cooling*). Adiabatic cooling occurs at a predictable rate because it is tied to atmospheric pressure. This rate is the dry adiabatic rate (DAR):

$$\text{DAR} = -\frac{5.5° \text{ F}}{1000 \text{ ft.}} \tag{1}$$

Along with the dry rate, unsaturated air has a dew point lapse rate (DP):[1]

$$\text{DP} = -\frac{1.1° \text{ F}}{1000 \text{ ft.}} \tag{2}$$

The dew point rate accounts for the reduction in dew point with increasing elevation. Atmospheric pressure decreases with elevation, reducing the force on water vapor to condense, which results in saturation occurring at lower temperatures even with the same humidity.

If $T(e) > D(e)$, then equations (1) and (2) explain the effect of elevation changes on T and D. Because DAR < DP, temperature is falling faster with elevation than dew point is dropping. Over normal ranges, there is a lowest elevation at which T and D are equal.

---

[1] These relationships are very general approximations of rates that actually vary with the vapor content of the air.

We call this the level of condensation, and represent this elevation by $\hat{e}$. When the level of condensation is reached, the air is saturated, causing temperature changes with elevation to be governed by the moist adiabatic rate (MAR):[1]

$$\text{MAR} = -\frac{2.5^\circ\,\text{F}}{1000\,\text{ft.}} \tag{3}$$

After saturation is reached, further cooling with elevation does not change the relative humidity, which remains at 100%. The MAR is less than the DAR because condensing water gives off latent heat. As ascending saturated air continues to cool, its capacity to hold water vapor decreases, causing excess water vapor to condense and precipitate.

Our model simplifies reality; the process is actually much more complex. For example, it is not necessary to have 100% relative humidity for precipitation to occur. These simplifying assumptions are necessary not only for the model, but are also used in our qualitative explanation of orographic precipitation.

2.3 Mathematical Models

2.3.1 Variables
We recall that:

e = elevation above the windward base in feet.

$D(e)$ = dew point in $^\circ$F at elevation e on the windward side,

$T(e)$ = air temperature in $^\circ$F at elevation e on the windward side,

$d(e)$ = dew point in $^\circ$F at elevation e on the leeward side,

$t(e)$ = air temperature in $^\circ$F at elevation e on the leeward side, and

$\hat{e}$ = level of condensation, the elevation at which condensation occurs.

2.3.2 Assumptions
We recall that:

1. If $T(e) \leq D(e)$, then precipitation occurs, with rain occuring if $T(e) \geq 32^\circ$ F, and snow occuring if $T(e) < 32^\circ$ F.
2. Temperature and dew point are in a straight line relation to elevation. In order to specify this relationship, we only need an initial condition and the rate of change.
3. We assume that the rates at which the temperature and the dew point change with elevation are given by equations (1), (2), and (3).
4. The maximum value of e (denoted M) and the minimum value of e (denoted m) are known (in feet).
5. We assume that we have measured $T(0)$ and $D(0)$ as our initial conditions at the base of the mountain on the windward side of the mountain, and that $T(0) > D(0)$.

2.3.3   Straight Line Model

On the windward side of the mountain, our initial conditions, rate of change, and assumed straight line relation give us, for $e \leq M$,

$$T(e) = \begin{cases} T(0) - 0.0055e, & \text{if } e \leq \hat{e} \\ T(\hat{e}) - 0.0025(e - \hat{e}), & \text{if } e > \hat{e} \end{cases} \tag{4}$$

and

$$D(e) = \begin{cases} D(0) - 0.0011e, & \text{if } e \leq \hat{e} \\ D(\hat{e}) - 0.0025(e - \hat{e}), & \text{if } e > \hat{e} \end{cases} \tag{5}$$

On the leeward side of the mountain, for $m \leq e \leq M$,

$$t(e) = T(M) + 0.0055(M - e)$$

$$d(e) = D(M) + 0.0011(M - e).$$

In order to apply this model, we need to know $\hat{e}$. We know that temperature and dew point are equal at the elevation $\hat{e}$, or in the notation of the model:

$$T(\hat{e}) = D(\hat{e}).$$

Applying (4) and (5), we solve for $\hat{e}$ to get:

$$T(0) - 0.0055\hat{e} = D(0) - 0.0011\hat{e},$$

or

$$\hat{e} = 227.\overline{27}[T(0) - D(0)].$$

We can now use the model to predict the temperature at any point on the mountain. For an elevation, e, on the windward side of the mountain, we can calculate $T(e)$ and $D(e)$ directly. For an elevation, e, on the leeward side of the mountain, we must first calculate $T(M)$ and $D(M)$. Rather than hand-calculate each temperature individually, we will automate the process with a spreadsheet or other computer software using a forecast model.

2.3.4   Forecast Model

We will maintain the same assumptions and variables, but cast the model as an algorithm that we can program a calculator or computer to run.

Identification of variables for mountain problem

$e =$ current elevation in feet above sea level

$M =$ maximum elevation in feet above sea level (mountain summit)

$m =$ minimum elevation in feet above sea level (valley floor)

$T(e) =$ Temperature in $^\circ$F at current elevation (windward side)

$D(e)$ = Dew point in $°$ F at current elevation (windward side)

Δe = standard elevation increment in feet as one moves up the mountain

h = next elevation increment in feet (used to keep e between M and m)

Pseudocode for naive algorithm (Misses level of condensation)

{Initialization}
Assign to e the value of zero
Get the values of M and m
Get the values of $T(e)$ and $D(e)$
Get the value of Δe
Write out the values of e, $T(e)$, and $D(e)$

{Move up the mountain}
While $e < M$ do
   Assign to h the value of the minimum of $\{\Delta e, M - e\}$
   {This value of h keeps $e \le M$ to avoid overshooting top of the mountain.}
   Assign to e the value of $e + h$
   {We may have just skipped over the level of condensation.}
   If $T(e - h) > D(e - h)$ then
     Assign to $T(e)$ the value of $T(e - h) - (5.5/1000)(h)$
     Assign to $D(e)$ the value of $D(e - h) - (1.1/1000)(h)$
   else
     Assign to $T(e)$ the value of $T(e - h) - (2.5/1000)(h)$
     Assign to $D(e)$ the value of $T(e - h) - (2.5/1000)(h)$
   Write out the values of e, $T(e)$, and $D(e)$

This algorithm works nicely in a spreadsheet, where graphing tools are readily available. This algorithm has one serious weakness; it is naive in the sense that it fails to change the rates at which temperature and dew point are dropping until an elevation above the level of condensation. The result is that temperature actually falls below dew point, an event that we have assumed does not occur. We have left the correction of this flaw as Exercise 3, with a suggested technique.

EXERCISES

1. *Applying the Model* Using the forecast model with initial conditions of

$$T(0) = 75° \text{ F} \quad \text{and} \quad D(0) = 55° \text{ F},$$

calculate $\hat{e}$ and complete the following table describing the weather conditions for each of the elevations given. (Assume a maximum elevation of 12,000 feet.) We suggest you employ a spreadsheet, computer program, or graphing calculator to implement the algorithm.

Elevation (in feet) on windward side	Temperature (in ° F.)	Dew Point (in ° F.)	Precipitation (yes/no/snow)
0	75	55	no
2,000			
4,000			
6,000			
8,000			
10,000			
12,000			

2. *Graphic vs. Algebraic Solutions* Solve the straight line model both graphically and algebraically. What advantages does each method have? What disadvantages?

3. *Spreading the Forecast Model* How can the naive algorithm be modified to implement the process of moving down the mountain? Be sure to employ a technique to avoid overshooting the base of the mountain.

Earlier, we identified a step in our naive algorithm where we will probably skip over the level of condensation. Can the algorithm be improved so that in moving up the mountain one does not skip over the level of condensation? Consider using the calculated value of the level of condensation, $\hat{e}$, and a technique similar to the technique used to avoid overshooting the top of the mountain.

4. *Leeward Side* Using the forecast model with initial conditions of

$$T(0) = 75° \, \text{F} \quad \text{and} \quad D(0) = 55° \, \text{F}$$

and the table completed earlier (see Exercise 1), complete the following table describing weather conditions on the leeward side for each of the elevations given. (Assume a maximum elevation of 12,000 feet.)

Elevation (in feet) on windward side	Temperature (in ° F)	Dew Point (in ° F)	Precipitation (yes/no/snow)
12,000			
10,000			
8,000			
6,000			
4,000			
2,000			
0			

Also try various values for M, m, $T(0)$, and $D(0)$. These values can be read from the internet, a daily paper, or television news program. Meteorological conditions make springtime data best for this exercise.

Why is there a difference in temperature, dew point, and precipitation between the windward and leeward sides at the same elevation? How much of an influence on $t(m)$ and $d(m)$ do M, $T(0)$, $D(0)$, and m have?